6-4-18
cond noted
OR

MAP 2015

What
Einstein
Didn't Know

What
Einstein
Didn't Know

Scientific Answers to Everyday Questions

15th Anniversary Edition
Completely Revised and Expanded

By Robert L. Wolke

DOVER PUBLICATIONS, INC.
Mineola, New York

Copyright

Copyright © 1999, 2014 by Robert L. Wolke
All rights reserved.

Illustrated by Diana Zourelias

Bibliographical Note

This Dover edition, first published in 2014, is a revised and expanded edition of the work originally published by Bantam Doubleday Dell Publishing Group, Inc., New York, in 1999.

International Standard Book Number

ISBN-13: 978-0-486-49289-6
ISBN-10: 0-486-49289-3

Manufactured in the United States by Courier Corporation
49289302   2014
www.doverpublications.com

To my daughter Leslie,
who inspired my teaching by continually asking,
"Why, Daddy?"

and to my wife, Marlene,
for her love, support, and encouragement.

Contents

3 *Around the House*   48

Explanations of everyday household mysteries, including: How
does soap know what's dirt? Why don't solids actually burn? How
does laundry bleach tell white from colors? Why do soda and beer
go flat? How can the same thermos container keep hot things hot
and cold things cold? Why do water beds need heaters? How do
batteries make electricity, and why do they die? Why won't your
shower temperature stay the way you set it? Why is it so hard to
find an object that you've just dropped on the floor? Is the five-
second rule valid?

4 *The Infernal Combustion Engine*   83

Solutions to automotive puzzles, including: Does your garage
floor suck the "juice" out of your battery? Why does straight
antifreeze freeze faster than a fifty-fifty mixture with water?
How does steel rust? Why won't sand provide traction for your
tires on ice? Does salt really melt ice? Why won't oil and water
mix? Why is oil such a good lubricant? How does carbon monoxide
kill? Why is the air from my tires so cold?

5 *The Great Outdoors*   104

Earth, sun, and air, including: Why is there always a cool breeze
at the seashore? Why do ocean waves always roll in parallel to
the shoreline? Why is the sun hotter at noon? Why is the winter
(in the northern hemisphere) colder than the summer? Why can
we see through air? Why is the barometric pressure measured
in inches? How can you tell the temperature by listening to a
cricket? How does the greenhouse effect work? Why is the Statue
of Liberty green? What eventually happens to a helium-filled
balloon after you let it go outdoors?

Introduction

This book explains what is going on behind everyday substances and occurrences. It is for people who are curious about the world around them, but who don't have the time or patience to search for explanations.

Here, you will not find facile, pop-science answers that would leave you just as mystified as before. Instead of mere answers, you will find plain-talk *explanations* that I hope will take you to a true understanding.

Explanations? Yes. I once wrote a college textbook called *Chemistry Explained*. (It is available for download at www.robertwolke. com.) I chose that title because I believe any science book must not just relate scientific principles, but explain them clearly to the reader until he or she understands them. I hope I have succeeded in reaching that goal in this book.

§

Traditionally, people have encountered science in one or more of four places: classrooms, textbooks, children's books, and serious tomes by solemn scientists. Unfortunately, classrooms and textbooks have turned off at least as many people as they have turned on. The fun books for kids are great, but they promote the false notion that curiosity ends at puberty. And those solemn scientific tomes only perpetuate the conviction that science is inherently incomprehensible to ordinary mortals.

This is not a textbook, it is far from solemn, and it is not a fun book for kids. It is a fun book for grown-ups. (But don't be surprised if your kids steal it from you.) Nor is it a collection of "science facts." Facts are not explanations. Instead, this book answers questions about the everyday wonders occurring around the house, in the kitchen, in the garage, in the marketplace, and in the great outdoors, even to the bounds of the Universe. There is no need to read this book sequentially. Browse to your heart's content and drop into any question that catches your eye; every explanation is self-contained. However, whenever some closely relevant information exists elsewhere in the book, you will be referred to the page on which it is explained.

As you browse, you will see a number of Try Its: experiments and demonstrations that you can do yourself, whether seated at your kitchen table or even on an airplane. You will also find a number of Bar Bets that may or may not win you a round of drinks, but that will certainly spark lively discussions.

IMPORTANT: Technical buzzwords, indicated by italics, are usually defined as they are introduced. But if you should forget a meaning, you will probably find it in the list of Buzzwords at the end of the book. Feel free to flip to the buzzwords list any time you need a definition.

§

Now about the title, "What Einstein Didn't Know." When my publisher told me he wanted that title, I objected strenuously. "People will think I'm claiming to know more than Einstein did!"

Well, let me tell you the cold, hard truth about authors and publishers: What the publisher wants, the publisher gets. So I lost the title battle. But it got your attention, didn't it? And that's what titles are supposed to do. Maybe publishers do know a thing or two about books.

Just don't expect this book to be about Einstein.

ROBERT L. WOLKE

CHAPTER 1

In the Marketplace

From the street vendor to the glitziest mall, it's the same old battle out there: people selling *versus* people buying. The sellers always have the advantage, because they know exactly what it is they're selling, while the *emptors* must be in a constant state of *caveat*. In many cases, the buyer not only doesn't know what the product really is, but can't even get a good look at it through the fog of promotion, packaging, and pitch.

In this chapter, we will take a clear-eyed look at what some products really are, beneath their glossy surfaces. We will visit a supermarket, a hardware store, a drug store, and a restaurant, with a stop or two at the local pub. I'll even throw in a plug for the metric system.

A Foggy Day at the Bar

I must have opened thousands of bottles of beer. (No remarks, please; I'm a bartender.) Many times, as soon as I pop the cap, wisps of fog appear in the neck of the bottle, and they sometimes even puff up above the opening. I've seen my share of foggy customers, but what causes foggy beer?

The fog is exactly the same as any fog: a collection of tiny particles of liquid water that have been condensed out of the air by a cold temperature, but that are too tiny to fall down like rain; they are kept suspended by being constantly bombarded by air molecules. They look white because they reflect all wavelengths of light equally.

Your puzzlement apparently stems from the fact that you can't see any fog inside the bottle until you open it, yet it is equally cold at both times. What is there about opening the bottle that makes the fog appear?

The space above the beer in the unopened bottle is filled with a mixture of compressed carbon dioxide, air, and water vapor—all gases. The water molecules in the vapor are content to stay that way—far apart from one another as an invisible gas, rather than clumping together as molecules of liquid—because they got there in the first place simply by leaping individually out of the beer's surface. At the temperature of the beer, only a certain number of them will have had enough vigor to leap into the void as vapor. Another way of saying this is that the amount of water vapor in the air above the liquid is in *equilibrium* with the liquid water at that temperature. Those vapor molecules remain suspended individually as a vapor until you remove the cap and release the gas pressure.

When you release the pressure, the compressed gases are able suddenly to expand, and when gases expand, they lose some of their energy and cool down (p. 98). The water vapor is now cold enough to condense into droplets of liquid water, and that's the fog you see.

Then, if you put the bottle down on the bar without pouring it, you—and your observant customer—may see some of the fog actually rising above the mouth of the bottle and spilling over onto the bar. With the pressure now released, carbon dioxide gas is leaving the beer and expanding as it hits the warmer air at the top of the bottle. As it expands, it lifts some of the fog above the bottle's rim. Since carbon dioxide is heavier than air, it actually spills over like an invisible waterfall, carrying some of the fog along with it and flowing down the sides of the bottle.

Now see if you can explain all this to the guy who asks, "Hey, man. Why's my beer smokin'?" Alternatively, keep a few copies of this book on hand and let him read it. Good for a vigorous discussion.

And no offense, but if you worked in a higher-class establishment, you would notice exactly the same fog effect upon opening a bottle of champagne, and for exactly the same reasons.

Pressing Hot and Cold

When I sprained my ankle playing softball, somebody ran to a drug store and bought a cold pack. They squeezed it and shook it, whereupon it turned into an instant cold compress. What's inside that package that makes it get cold so fast?

The cold pack contains ammonium nitrate crystals and a thin, breakable pouch of water. When the pack is squeezed, the water pouch breaks and, with a little shaking, the ammonium nitrate dissolves in the water.

When any chemical dissolves in water, it may either absorb heat—get cold—or release heat—get hot. Ammonium nitrate is one of those that absorb heat. It takes the heat right out of the water, thereby cooling it. And the amount of cooling is not trivial; that cold pack can actually get down close to freezing.

Because doctors keep blowing hot and cold about when to apply heat to an injury and when to apply cold, there are almost as many hot packs on the market as there are cold packs. The hot packs contain one of those chemicals that give off heat when they dissolve in water, usually crystals of calcium chloride or magnesium sulfate.

But why should a chemical absorb or release heat during the simple process of dissolving in water? After all, at home we dissolve crystals of two common chemicals, salt and sugar, in water time after time, yet we never see the sugar, for example, cooling off our hot coffee or heating up our iced tea. The fact is that salt and sugar are exceptions (see below).

When a chemical substance dissolves in water, it is a two-step process: first, the chemical's solid, crystalline structure must be broken down, and then a reaction takes place between the water and the broken-down chemical parts. The first step invariably has a cooling effect, while the second step has a heating effect (see below). If step one cools more than step two heats, as in the case of ammonium nitrate, the overall effect is cooling. If it's the other way around, as it is with calcium chloride and magnesium sulfate, the overall effect is heating. In the cases of salt and sugar, the two steps happen to be just about equal, so they cancel each other out and there is very little change in temperature.

Here is what's going on during the two-step process in which a solid crystal dissolves in water. FYI, a crystal is a rigid, three-dimensional, geometric arrangement of particles. The particles may be atoms, ions (charged atoms), or molecules, depending on the substance we're talking about; we'll just call them particles.

Step 1: First, the particles must be released from their rigid positions in the crystal in order to be able to float about freely in the water. To break down any rigid structure requires the expenditure of energy; somebody or something has to supply the sledgehammer blows that knock the structure apart. During the breakdown of the crystal's structure, therefore, some heat energy must be borrowed from the water, and the water cools down accordingly.

Step 2: The liberated particles don't just swim around in splendid isolation. They have a strong mutual attraction for water molecules. If they didn't, they wouldn't have been interested in dissolving in the first place. As soon as they are in the drink, they are literally attacked by water molecules, rushing to cluster around them like magnetic mines around a submarine. When magnets (or water molecules) are attracted to something, they expend energy in their rush toward their targets. This energy, the *energy of hydration*, heats up the water.

Now it's just a matter of which effect is bigger: the cooling effect from the breakdown of the solid or the warming effect from the particles' attraction for water molecules. If the cooling is bigger, the net effect will be that the water gets colder when the solid dissolves. That's how it is with ammonium nitrate. On the other hand, if the warming effect is bigger, the net result is that the water gets warmer when the solid dissolves; that's how it is with calcium chloride and magnesium sulfate.

Salt and sugar? In each case, it's just an accident that the two effects are approximately equal and cancel each other out. So there is practically no net cooling or heating when salt or sugar dissolves in water. (Actually, salt—sodium chloride—does cool the water very slightly when it dissolves.)

TRY IT   *Ammonium nitrate is a common fertilizer and calcium chloride is a common dehumidifier, sold for drying out damp closets and basements. You may have some of these chemicals around the house or farm. Stir some ammonium nitrate into water and the water will get very cold. Stir some calcium chloride into water and it will get quite hot. (Don't cover and shake; the heat can make the liquid splatter.) A tablespoon of the solid in a glass of water will do.*

Oysters on a Half-Shelf

Half the calcium supplements on the health-food shelves seem to be ground-up "natural oyster shell." Is oyster-shell calcium better than other kinds?

Clams and oysters make their shells primarily out of calcium carbonate. But chemically speaking, it doesn't matter whether the calcium carbonate in the supplement bottle came from an oyster bed or a bed of limestone, which is also made of calcium carbonate. Neither is more "natural" (whatever that means) than the other. Calcium carbonate is calcium carbonate. Oysters incorporate a bit of non-mineral matter in their shells, however, so calcium carbonate from other sources might be a bit purer.

Calcium supplements are sold in other chemical forms besides calcium carbonate (read the labels). Weight for weight, though, these other forms contain less calcium than calcium carbonate does, and it's the actual element calcium that you're after; your metabolism doesn't care about the other stuff. Calcium carbonate contains 40 percent calcium by weight, while calcium citrate contains 21 percent, calcium lactate contains 13 percent, and calcium gluconate contains only nine percent calcium. Now you can figure out which supplement on the shelf gives you the most calcium for your money.

But bear in mind that different chemical forms of calcium may be absorbed to different degrees in different people's bodies. Nutritionists argue incessantly about this.

The Great Fog Forgery

Why is dry ice dry? And what makes all those clouds of smoke around it?

It's not smoke; it's fog. And it isn't carbon dioxide either, as some people think; carbon dioxide gas—CO_2—is invisible. The fog surrounding the dry ice is made of tiny droplets of water, condensed out of the air's natural humidity by the dry ice's low temperature.

Dry ice is carbon dioxide in solid form, just as regular ice is water in *its* solid form. Water ice cannot be heated beyond 32°F (0°C) without "desolidifying" (transfiguring from the solid to a different state), while dry ice cannot be heated beyond −109°F (−78.5°C) without desolidifying into—turning into—not *liquid* CO_2, but *gaseous* CO_2, because it cannot exist in liquid form at normal atmospheric pressure. So a chunk of dry ice that is desolidifying (if I use that often enough, it may become a real word) into a gas is much, much colder than a chunk of ordinary ice melting into a liquid. Regular ice is wet because as it melts it becomes liquid water. Dry ice is dry because it doesn't melt; it changes directly into a gas without becoming a liquid first.

Why doesn't CO_2 like to be in the liquid state?

Well, carbon dioxide molecules don't like one another very much; they don't stick together very well, the way water molecules do. Water

You didn't ask, but...
Why does a CO_2 fire extinguisher shoot out a blast of snow?

It's not water snow, but CO_2 snow: flakes of dry ice.

A CO_2 extinguisher is nothing but a high-pressure tank of liquid carbon dioxide with a squeeze valve. When you squeeze the valve, you let some of the liquid CO_2 inside the tank escape. It instantly becomes a blast of very cold CO_2 gas, accompanied by flakes of solid CO_2 and a fog of water, condensed from the air. The extinguisher works in two ways: the coldness of the vapor can lower the temperature of the fire's fuel below its ignition point, while the carbon dioxide smothers the fire because it's a heavy, non-flammable gas that pushes away the oxygen.

Dry ice has been used on movie sets to fake fog. It is real fog, all right, because it consists of microscopic droplets of water suspended in the air. But you can always tell a fog forgery, because the water is very cold from the dry ice and the fog therefore lies on the ground like a blanket—unless it is blown around by an off-camera fan or stirred up by a mob of stumbling zombies. Real, weather-generated fog, on the other hand, hangs fairly motionless in the air.

Movies and plays use dry ice also to simulate cauldrons of boiling water. Just throw some dry ice into the water, and as the solid carbon dioxide changes to gaseous carbon dioxide, it rises through the water in fog-filled bubbles that break at the surface and are supposed to emulate hot steam. If you look closely, though, you can always tell that it's fake. Steam goes straight up because hot air rises, while the cold dry-ice fog hangs low over the cauldron. Again, like a blanket.

While we're on the subject of movie fakes, how about those scenes of storm-tossed ships? Are they just miniature models, shot at slow motion in a big tank? There's a surefire way to tell. Check the size of the water droplets from the crashing waves. If they're the size of a porthole or a cannon ball on the ship, it's a model in a tank. Water just doesn't break up into drops the size of cannon balls, unless the "cannon balls" on the ship are really BBs on a scale model.

molecules, H_2O, have a central oxygen atom with two hydrogen atoms sticking out like devils' horns at an angle of 104.5 degrees to each other. In liquid water, those hydrogen horns form weak *hydrogen bonds* between adjacent molecules, binding them together with a mild sort of stickiness. This hydrogen-bond stickiness between water molecules is responsible for a number of unusual properties that make this common liquid categorically unique (p. 94).

But CO_2 molecules are shaped like this: O=C=O. They have no sticky hydrogen horns to bind them together, so they cannot condense into the tightly crowded structure of a liquid unless forced together by a high pressure. Carbon dioxide is shipped around the country this way—as a liquid under high pressure in steel tanks. When the tank's valve is opened, the liquid instantly boils off into a burst of gas.

You didn't ask this either, but . . .
Why is the blast of snowy gas that comes out of a CO_2 fire extinguisher so cold, even though the extinguisher may have been sitting around in the room for months?

When you release some of the pressurized, liquid carbon dioxide from the high-pressure tank into the low-pressure room, it instantly flashes off into a gas, which must then expand rapidly under the reduced pressure. In order to expand, the gas must make room for itself by knocking other stuff, say the air molecules in the room, out of the way. When the CO_2 molecules knock the air molecules for a loop, they lose some energy and slow down, just as a billiard ball loses some energy when it collides with another ball. And molecules that have slowed down are lower-temperature molecules by definition—low enough in some parts of the cloud to freeze into flakes of dry-ice snow.

This is not just a phenomenon of carbon dioxide. Any gas, when expanding (except when expanding into a vacuum), will lose energy and cool down. We will see this situation again when we let some air out of our overinflated car tires (p. 98).

My True Love Is Nothing but Skin and Bones

Somebody tried to tell me that clear, shimmering, sparkling-bright Jell-O, my childhood true love, is made from pig skins, cow hides, bones, and hoofs. Yuck! Can that possibly be true?

Of course not. Just the skins and bones. No hoofs.

Jell-O and similar desserts are about 87 percent sugar and 9 or 10 percent gelatin, plus flavoring and coloring. Kids love three things about it: it is brightly colored, it is very sweet, and it jiggles. (Love that jiggle.) Mothers don't mind, because gelatin is pure protein.

The gelatin, which of course is the jiggler, really does come from pig skins, cattle hides, and cattle bones. But stop squirming. Every time you have made soup or stock that jelled in the refrigerator, you have made gelatin out of chicken hides or beef bones.

The skin, bones, and connective tissue of vertebrate animals contain a fibrous protein called collagen. There is no collagen in hoofs, hair, or horns. When treated with hot acid (usually hydrochloric or sulfuric acid) or alkali (usually lime), collagen turns into gelatin, a somewhat different protein that dissolves in water. The gelatin is then extracted into hot water, boiled down, and purified.

Purified? You bet. But you don't want to see (or smell) the early stages of the process. By the time the gelatin leaves the factory, it has been thoroughly washed at various stages to get the acid or alkali out, then finally filtered, deionized (a way of removing chemical impurities), and sterilized. What eventually leaves the factory is a pale yellow, brittle, plastic-like solid in the form of ribbons, noodles, sheets, or powder. When solid gelatin is soaked in cold water, it absorbs water and swells up; then, when heated, it dissolves to form a thick liquid, which jells on cooling.

As a protein, gelatin is obviously nutritious, although it is not what nutritionists call a complete protein. What is most fun about it is that when dissolved in water it's a gel (a jelly-like substance) when cold and a liquid when warm. It literally melts in your mouth. That's the main characteristic that it gives to confections from marshmallows to Gummi Bears, whose gumminess comes from a high proportion, around 8 or 9 percent, of gelatin. And guess what holds those tiny white dots on the tops of nonpareil chocolate candies? Right. Gelatin, used as an adhesive.

Most of the gelatin made in the US—more than a hundred million pounds a year—is slurped in the form of gelatin desserts. You'll find it also in soups, shakes, fruit drinks, canned hams, dairy products, frozen foods, and bakery fillings and icings. However, food isn't the only use for this unique substance. Those little two-piece capsules that many drugs come in are made of gelatin—about 30 percent gelatin in 65 percent water. Gelatin is so ubiquitous, in fact, that kosher-observing Jews and halal-observing Muslims have to be assured that no pig parts went into the cauldron that the gelatin was made in.

And then there is photography. Remember photographic film? Before digital cameras? The photographic emulsion—that thin, light-sensitive coating on the film—was made of dried gelatin containing the light-sensitive chemicals. Nothing better than gelatin has been found for that purpose since it was first used in photography in 1870.

The Proof Is in the Drinking

Wine and whiskey labels state the alcoholic strength as "proof" or as "percent alcohol by volume." Where does "proof" come from, and what does "by volume" mean?

The term *proof* was coined in the 18th century when the British *proved* (in its earlier meaning of "tested") the alcohol content of their sailors' rum ration by moistening gunpowder with it and setting it afire. Yes, gunpowder. Rum and gunpowder were two readily available commodities on British warships.

If the rum contained enough alcohol, more than about 57 percent by volume, the alcohol vapors would ignite and kindle the gunpowder. (The gunpowder would burn, but not explode, because it was not confined in a sealed space, such as in a bullet casing or a pipe bomb.) That 57-percent concentration of alcohol was defined as 100° (100 degrees) proof. If the rum was weaker than that, its vapors would not ignite and set the gunpowder afire.

In the US, 100 proof is defined as 50 percent alcohol by volume, so the proof is always twice the percentage of alcohol. Most booze today is bottled at 80 proof, or 40 percent alcohol by volume.

Note that alcoholic strengths are always quoted as percent of *alcohol by volume* (ABV), not as percent by weight. They are not the same. If you mix equal *volumes* of alcohol and water, the resulting mixture will be significantly different from a mixture of equal *weights* of alcohol and water. That may sound odd, but alcohol and water do behave oddly. There are two reasons for this.

• First of all, alcohol is lighter (less dense) than water. A pint of pure alcohol weighs only 79 percent as much as a pint of water.

Let's say we want to make a 50-percent-by-weight mixture of alcohol and water by weighing out equal weights—pounds or grams—of these two liquids and mixing them together. We'd find that we have to use a bigger volume of alcohol to get the same weight as the water. By weight, the mixture will certainly be 50 percent alcohol, but by volume it will be *more than* 50 percent alcohol. It works out to be about 63 percent.

Now guess which type of percentage the beverage manufacturers have chosen to quote on their labels. Right. The one that makes the alcoholic content seem higher: percentage by volume. Taxes are usually based on the percentage of alcohol, so the tax man also profits by this dodge.

• The second reason that percentage by volume is the chosen measure for wines and liquors is that a very unusual thing happens when alcohol and water are mixed: the final mixture takes up less space than the sum of the volumes that were mixed. In other words, the liquids shrink. Mix a pint of alcohol with a pint of water, and you get only 1.93 pints of mixture instead of the two pints you would expect. The reason is that water molecules and alcohol molecules form *hydrogen bonds* to each other, which snuggles them together even more tightly than they were in the pure alcohol and pure water themselves.

As you can imagine, this screws up the concept of percentage of alcohol by volume. Should it be the percentage of the volumes before mixing, or the percentage of the final volume after mixing? The beverage producers have decided to use the smaller volume, the volume after mixing. That is quite appropriate, of course, because that's the way we buy the product: already mixed.

If you are not too deeply mired among the mathematically challenged, you will quickly realize that this method of reckoning gives an even higher value for the percentage of alcohol. Score another one for marketing. Using this calculation method, that 50-percent-by-weight mixture we so carefully weighed out appears on the label as that 63 percent by volume we saw above.

Bar Bet. I can mix a pint of alcohol with a pint of water, and get a mixture that is more than 50 percent alcohol by volume.

The Old Booze Bottle Ain't What She Used to Be

How have liquor bottles changed over the years?

Before January 1, 1980, distilled alcoholic beverages in the US were sold in bottles containing either a quart or a fifth of the liquor. A *fifth* was defined as one-fifth of a gallon. Because there are four quarts in a gallon, a fifth was equal to four-fifths of a quart.

In the 1970s, agitation was growing for the US to switch its measurement units to the metric system (see p. 191), in order to be in step with the rest of the world. The years to date have seen absolutely no movement in that direction (three cheers for progress!), with only two noticeable exceptions: soda pop now comes in two-liter bottles, and federal law—US Code of Federal Regulations Title 27 › Chapter I › Subchapter A › Part 5 › Subpart E › Section 5.47a, effective January 1, 1980—has replaced the old fifth with a new 750-milliliter size.

If you do the math, you will find that the old fifth was equivalent to 757 milliliters, seven milliliters more than the new metric size. Because the bottles looked very much the same, hardly anyone seemed to notice the switch. But did the distillers lower their prices by seven 757ths, 0.925 percent or about a penny on the dollar?

Not on your life. We consumers had to swallow the difference, so to speak.

How Is It Maid?

How come all my Rubbermaid household products are not "maid" of rubber? They're plastic. And why don't they spell it Rubbermade or Plasticmade or, especially these days, Chinamade?

It all began in 1920 in Wooster, Ohio, where there was a company called Wooster Rubber. They made toy balloons and sold them under the brand name Sunshine.

Before we had rubber to make things out of, many ancient civilizations blew up the urinary bladders of animals, usually pigs, tied them off, and played games with them. Pigs' bladders are thin, lightweight, airtight, and elastic—perfect to be thrown, kicked, or batted around.

On Christopher Columbus's second voyage to the New World, he found Indians on Haiti playing with balls that bounced much higher than his crew's bladders—the ones they brought with them for recreation, that is. The Indians were using balls made from a thick, milky liquid that oozed from certain tropical trees. They called it *caoutchouc*; we know it as *latex*.

Most natural rubber—as opposed to the synthetic rubber that was devised by chemists during World War II when natural sources were unavailable—comes from a single species of tree, *Hevea brasiliensis*, which is native to South America. Today, *H. brasiliensis* is grown in large plantations in Southeast Asia, especially in Malaysia.

In 1736, a French scientist named Charles Marie de la Condamine, sent by his government to do astronomical research in Ecuador, brought back some samples of *caoutchouc*. When it was discovered in 1770 that pieces of *caoutchouc* rubbed out pencil marks from paper, it became known as rubber or India rubber, because it came from the Indies. Small rubber erasers became popular in France as *peaux de negres* (Negro skin) because of their dark brown or black color. Throughout Europe, rubber erasers quickly replaced stale bread as the accepted material for rubbing out pencil marks.

Around 1862, an English leatherworker began replacing the pigs' bladders inside rugby footballs with rubber, and just as Columbus found out in Haiti, they bounced better. (A cartoon in a recent issue of the *New Yorker* shows primitive Haitian Indians at their traditional game of kicking around a severed human head, when another Indian comes by with a ball made of rubber. One of the first group of Indians says to the intruder, "I don't care if it's more bouncy. It threatens the integrity of the game.")

You didn't ask, but . . .
How do they make those very thin rubber goods like rubber gloves and toy balloons?

Surgical gloves and recreational products such as balloons and condoms are made by dipping appropriately shaped solid molds into vats of liquid latex: *caoutchouc* mixed with water and some ammonia to keep it stable.

The latex coats the mold in a thin film and dries to a sheath of the intended shape, which is then stripped or rolled down off the mold. For toy balloons, the mold is shaped like a duckpin. For condoms, the mold is—no, it's not what you think—a glass tube with a rounded end.

Now back to Rubbermaid. In 1933, a Connecticut couple named Ralph and Madeline Caldwell decided there would be a market for household implements made of colored rubber. They made and patented a bright red rubber dustpan and promoted it to housewives as "your rubber maid." Other rubber products followed under the Rubbermaid name. Their success caught the eye of one of the owners of the Depression-wracked Wooster Rubber Company and a merger was arranged under the Rubbermaid name. The new company, now called Newell Rubbermaid, introduced its first plastic product in 1956: a plastic dishpan. And they were off and running in the international plastics sweepstakes.

Pumping Iron

I've been told that some breakfast cereals, fortified with minerals and vitamins, actually contain pieces of iron metal. Can that be true?

Surprisingly, yes. They aren't the size of railroad spikes, but they are indeed pieces of metallic iron. Tiny, almost microscopic pieces.

Iron is considered one of the essential minerals—that is, *chemical elements*—required in a healthy human diet. Needed in only tiny amounts, it plays many vital roles in our bodies. One of the most important is that it makes up part of the hemoglobin molecule, which transports vital oxygen to all our cells via the bloodstream. Iron is also part of the myoglobin molecule, which stores oxygen in our muscles for instant availability.

The US RDA (Recommended Dietary Allowances) of iron is eight milligrams per day for adult males and 18 milligrams per day for premenopausal females. That's not much iron; a single staple from an office desk stapler weighs about 36 milligrams.

How do they fortify cereals with iron? There are many chemical compounds of iron that would do the trick, but nontoxic ones are, of course, highly preferable. (Iron toxicity doesn't begin until doses higher than about 700 milligrams for a 150-pound adult.) Compounds such as ferric phosphate and ferrous sulfate are often used to fortify flour, breads, and cereals.

But the actual amount of iron in these compounds constitutes only 37 percent of their weight. So why not add 100-percent iron directly? Finely divided metallic iron is much cheaper than those and other compounds, and you know that "cheap" is a primary motivation for food manufacturers. So they add pure iron metal and call it reduced iron, because it has been obtained by a process called *chemical reduction*. Reduction is the opposite of *oxidation*, the process by which iron rusts.

The metallic iron in your fortified cereal is not in the form of filings, as some "natural health" gurus claim in their attempts to damn it as "unnatural." It's actually in the form of an extremely fine *powder* of particles about five microns, or about two ten-thousandths of an inch, in size. When they fall into the hydrochloric acid in your stomach, they dissolve in a flash and the iron atoms (actually, ions) are free to be absorbed and metabolized by your body.

Because some of this metal winds up in the hemoglobin of your bloodstream, you would be perfectly justified in claiming that your heart is "pumping iron."

The Shot Pulled 'Round the World

Ever since Starbucks conquered the world, espresso coffee is everywhere. Almost every restaurant has it, with varying degrees of quality. Sometimes it has a tan or brown foam on the surface and sometimes it doesn't. What's the significance of that foam, and why doesn't all espresso have it?

Starbucks was founded in Seattle in 1971 and began expanding worldwide six years later. As of this writing, there are 20,891 Starbucks emporiums in 62 countries. Whoops! Make that 63 countries. Correction: 64. No, 65. Well, you get the idea. They've been popping up like 17-year cicadas. So if you ever need a caffeine fix in the city of Bacolod in the province of Negros Occidental in the Western Visayas islands, have no fear. There is a *Quad Venti Skinny with Whip Iced Caramel Macchiato* with your name on it just around the corner.

Most of Starbucks's bizarrely named concoctions have a base of espresso, which is a complex emulsion of water-insoluble oils, with sugars, acids, proteins, and flavors extracted by hot water, forced under pressure through the finely ground coffee beans. The pressure can come from either steam or a manual pump, depending on the design of the machine. "Pulling" a fine "shot" of espresso by an expert barista is an exacting science.

The foam, referred to by Italians as *crema*, is made of bubbles of carbon dioxide gas that are stabilized—similarly to how soap stabilizes bubbles—by compounds called *melanoidins* that are formed in the coffee when sugars and amino acids react with each other.

The carbon dioxide had been generated during the roasting of the coffee beans, and some of it remains even after the beans are ground. But if the ground coffee is left to stand around for several minutes before being hit with the pressurized hot water, much of that carbon dioxide will escape, killing the possibility of a good head of *crema*. That's why every cup of espresso must be made fresh, immediately after the beans are ground. Stale coffee that has lost its carbon dioxide will not form a *crema*, nor will an espresso machine that isn't scrupulously clean.

An expert barista knows exactly when to stop pulling the shot: too late, and many harsh, bitter flavors will be extracted; too soon and the brew will be too thin to support a robust head of crema. A good, stable crema is one sign of a carefully made, high-quality cup.

Just as with soap bubbles, the crema bubbles will break down as the cup stands around waiting for a slow drinker. So don't dawdle. And if your espresso arrives without any crema at all, just say, "*No, grazie*" and "*arrivederci!*"

Let There Be Lux!

A light bulb blew out in my reading lamp and I went to buy a new one. To my chagrin, I found an astounding number of different kinds! There were not only the familiar pear-shaped household incandescents, but lots of CFLs and LEDs, each in a profusion of wattages, sizes, and shapes. How can I pick the best one for my purpose?

The industrialized world is in the midst of a technological revolution, not just in the realm of computers and other electronic marvels, but in the simple quest for light. And the current awkward period will continue to bedevil us for some time to come.

From the day a few hundred thousand years ago when one of us humans discovered fire, almost up until the twentieth century, the only kind of light we could produce at will was firelight: by burning wood, candles, lanterns, and gas lamps.

Then, in 1880, Joseph Swan in England and Thomas Edison in the US patented their incandescent electric light bulbs, and we've been using them virtually unchanged ever since. In fact, Edison's original

27 mm-diameter, seven-threads-per-inch screw base is still used on bulbs in the US and most of Europe.

But incandescent light bulbs are incredibly inefficient. About 90 percent of the electric energy they use becomes heat, not light. Using those hot devices as a light source is about as efficient as trying to read by the light of a hot poker. But that's all we've had for more than a hundred years, until the light-bulb revolution began in the 21st century. Now that we know our planet is heating up because we've been burning so much fossil fuel, we can't waste energy that recklessly any more.

Countries around the world have been phasing out incandescent light bulbs in favor of more modern, more energy efficient alternatives. Brazil and Venezuela started their phaseout in 2005; the European Union, Switzerland, and Australia in 2009; Argentina, Russia, and Canada in 2012; and the US and Malaysia in 2014.

Enter CFLs and LEDs. CFLs (compact fluorescent lamps) are simply conventional fluorescent-light tubes that have been wound up into a spiral to save room. At each end of the tube is a tungsten electrode that emits electrons when heated. An alternating (AC) voltage pulls the electrons back and forth between the electrodes, reversing their direction fifty or sixty times a second. The tube is filled with a low pressure of gas (usually argon) plus some mercury vapor evaporated by the filaments' heat from tiny globules of liquid mercury.

When the alternating stream of electrons collides with a mercury atom in the gas, the atom absorbs some of the electron's energy (it becomes *excited*) and spits it back out as *ultraviolet* (UV) radiation. But UV is invisible to humans, so we have to convert it into light we can see. That job is done by the white coating (a *phosphor*) on the inside surface of the tube. It absorbs the ultraviolet light and re-emits it as visible light, a process called *fluorescence*. Because fluorescent lamps need electricity only to heat the two small filaments to a relatively low temperature, they use only about one-fourth the power that a comparable incandescent bulb would use to give the same output of light.

LEDs (light-emitting diodes) operate on a well-established principle of quantum theory: that an electron in an atom can exist with any one of several fixed amounts of energy (*energy states*), but no amounts in between. The permissible energy states are like the rungs of a ladder: you can't climb up or down by stepping on the spaces in between. When an electron in one of the higher energy levels falls down (in energy, not distance) to a lower one, it discards its surplus energy in the form of a *quantum*—a discrete bit—of electromagnetic radiation called a *photon*.

In an LED there are two chips of semiconductor crystals, one of which (an *N-type semiconductor*) contains its full complement of electrons, while the other (a *P-type semiconductor*) has been "doped" with impurities that make it deficient in electrons; it has "holes" where the electrons should be. The surfaces of the two crystals are in contact with each other at what is called a boundary, or *junction*. When such a bread-only sandwich, called a *diode*, is placed between two electric terminals, it allows current (electrons) to flow in only one direction: from the N-type into the P-type, where the electrons "fall into" the holes as if stepping into an open manhole, while emitting their energy as photons. The photons might be anything from infrared (low-energy) to ultraviolet (high-energy) light, depending on the amount of energy the electrons give up by falling into the holes.

The remote control for your television contains an LED that gives off invisible, low-energy infrared light in Morse-code-like blinks that tell the TV which button you pressed. LEDs made for household lighting are designed to emit photons of several colors that, when combined, appear to our eyes as white light. Today's commercial LEDs produce up to five times the amount of light per watt of electric power as an incandescent bulb. They are clearly the future of energy-saving illumination.

Okay, so how do you shop for light during this confusing transition period? All you have to do is compare the bulb's base with the size of the socket, because the standard Edison screw base has variants: the smaller Candelabra and the larger Mogul (in North America) or Goliath (in Europe). I'll assume you can screw in the light bulb without any help from a number of your countrymen.

Then, read the information on the package. It will tell you how much light, measured in *lumens*, the lamp will produce. Remember, what you're shopping for is *light*; nothing else really counts. The package will also tell you how long the lamp is likely to last and how much money you will save in electricity costs. Ignore those numbers; they are only estimates and sales gimmicks. Concentrate on the number of lumens. Recall how much light you used to get from a typical incandescent bulb, and buy approximately the same number of lumens: a 40-watt incandescent bulb gave about 500 lumens; a 60-watt, about 850 lumens; and a 100-watt, about 1700 lumens.

With those numbers in mind, pick whatever crazy-looking lamp you fancy that will give you the number of lumens you want.

The number of lumens a lamp produces is its total light output in all directions. But what's more important is the intensity of light that actually falls on the surface of your book's page. The *light intensity*, or *illumination*, is expressed in units of *lux*, abbreviated lx, a word that is both singular and plural. One lux is what used to be called a *foot-candle*.

The light intensity—the number of lux—falling on the page of this book you're reading depends on its distance from your lamp and on so many other factors that it cannot be specified for any given situation. Just pay attention to the number of lumens. Under any given circumstances, the more lumens a lamp puts out, the more lux you'll get on the page.

Lotsa lux to you.

Jimmie or Sook?

Just for fun, whenever I see live crabs at the fish market I ask whether they're males or females, and I'm always told that they're males. What's the reason for that?

Pure sexual discrimination. Being larger, the males are more suitable for steamin'-'n'-pickin', so they are sold screamin'-'n'-kickin'. The females have usually been spirited away to be dismantled at the wholesale level and packed as canned crab meat.

You can always tell the males because they have blue claws and they are the ones holding the TV remote. (Kidding.) The females have reddish claws, looking as if their "fingernails"—the tips of their claws—had been painted red. (Not kidding.)

Or, just look at the critter's underside and you will see an apron: a thin flap of shell that covers the rear of the abdomen. If the apron is shaped like the Eiffel Tower in Paris, it's a male or *jimmie*. If the apron is shaped like the Capitol dome in Washington, it's a mature female or *sook*. Nobody seems to know the origin of that word.

Who does all the segregating before consigning them to cauldron or can? The crabs themselves, because except when mating, they tend to wander off to different locations, like married couples at the mall. That's because the Chesapeake Bay, where most American blue crabs come from, is fed by fresh-water rivers from the north and by ocean tides from the south, so that its thousands of coves and inlets contain just about every degree of saltiness (*salinity*) that a crab's little heart could desire: saltier for hatching eggs, for example, and less salty for general horsing around. So at various stages in their lives the males and females hang out in different stomping grounds. Crafty Chesapeake watermen can go out and capture jimmies or sooks separately, as the market demands.

Superman Unglued

I think Super Glue is amazing. How does it stick objects together so incredibly quickly and strongly?

Super Glue is reputed to be so strong that only Superman could unglue it, and that's not much of an exaggeration. What kind of technological wizardry could come up with the formula for such a powerful chemical concoction? And what kept humans from inventing it until the middle of the twentieth century?

Believe it or not, the glue isn't a complex chemical composite; it's a pure chemical compound called *ethyl cyanoacrylate*. It is an *ester*, the product of the reaction between an acid and an alcohol. The general formula of this particular ester is $CH_2=C(CN)COOR$, where R stands for the alcohol it was made from: methyl (CH_3) for methyl alcohol, ethyl (C_2H_5) for ethyl alcohol, etc.

What makes cyanoacrylate esters so special among the hundreds of adhesives and millions of chemical compounds known to science and industry? Thereby hangs—tightly fastened, I presume—a tale.

During World War II, an Eastman Kodak Co. chemist named Harry Coover was trying to develop a clear plastic that could be used in military gun sights. Among the compounds he investigated were cyanoacrylates, but they were much too sticky to work with, and he abandoned them. In 1951, he was put in charge of a group trying to develop a strong, transparent, heat-resistant plastic for use in aircraft canopies, which were at the time being made of *poly(methyl methacrylate)*, commonly known as Plexiglas. Notice the "acrylate" in that chemical name. Acrylates are esters similar to cyanoacrylates, which Coover and his crew then turned to as possibilities.

One important property of a transparent optical material is its refractive index: the degree to which it can bend a light ray, as lenses do. To test its refractive index, one of Coover's colleagues put a couple of drops of ethyl cyanoacrylate between the two glass prisms of an instrument called a *refractometer*. It glued the prisms together so tightly that they couldn't be separated, and the expensive instrument was ruined. Instead of firing the culprit, one Fred Joyner (appropriate name for a person working with adhesives), Coover said, "Eureka! Aha!" or something like that, and superglue was born.

Cyanoacrylate adhesives were introduced commercially in the early 1970s and today are a billion-dollar industry, not because of those tiny squeeze-tubes you buy at the hardware store, but because they are extensively used as an adhesive in surgery and other medical procedures for sealing wounds or controlling bleeding.

Both poly(methyl methacrylate) and superglue cyanoacrylates are *polymers*: substances made up of hundreds, thousands, or even millions of repeated units of a small, simple molecule (called a *monomer*), all strung together into one giant molecule. Starch, cellulose, rubber, and DNA are natural polymers, while most plastics are man-made polymers.

What's unique about cyanoacrylates is that their small monomer molecules can *polymerize* (form a polymer) in a jiffy. They can spring into a linear formation faster than a bullying drill sergeant can line up recruits. Expose the liquid to any chemical compound containing a *hydroxyl group* (OH), such as water—and water is present on virtually all surfaces that are exposed to humidity in the air—and the monomer molecules will snap into a tough, stable polymer form almost instantly. If a thin film of the monomer is placed between two adjacent surfaces, the resulting polymer will bind them together quickly, and as strongly as a weld between metals.

This process is fundamentally different from how most glues and adhesives work. They are usually solutions of a sticky substance in a solvent such as water or an organic solvent. Elmer's white glue, for example, is a water solution of a rubber-like polyvinyl acetate (PVA) latex. Spread it over two surfaces, put them together, let the water evaporate (i.e., the glue dries) and the surfaces will remain bonded together by the latex.

Now for the bad news. If you're not very careful when working with a superglue, it is devilishly easy to glue your fingers together or glue a finger to the object you're working on, and it is devilishly difficult to get them apart. There is always some moisture on your skin that will initiate the polymerization process, while the skin's ridges and porousness will anchor it there. If you're stuck, try soaking the joint in warm water or better yet, apply nail polish remover. It contains acetone, which will degrade the polymer. Don't try to pull your finger away, or you may leave some skin behind.

Super Glue (capitalized) is the trademark of a leading brand of cyanoacrylate adhesive. Others are sold under the names Krazy Glue, Loctite, Gorilla Glue, and many more.

CHAPTER 2

In the Kitchen

There is no place in our daily lives where so many marvelously mysterious things are going on as in the kitchen. That's where we mix, heat, cool, freeze, thaw, and occasionally burn an incredible assortment of animal, vegetable, and mineral materials, using equipment that would have rattled the retorts and curdled the cauldrons of the most devoted alchemist. It's no accident that Shakespeare chose "Fire burn and cauldron bubble" as the most mystical manipulations of his witches in *Macbeth*.

Beneath the surface of these familiar operations, some extraordinary transformations are taking place, transformations of the sort that the alchemists could only have fantasized about, but that we can now explain in the simplest of terms.

Do you think you already know what's going on when you boil a pot of water? Think again. We're going to start by taking a close look into that very pot to see what makes the witches' cauldron bubble.

What's the Point of Boiling?

When I put a pot of water on the stove, I turn up the heat as high as it will go because I'm always in a hurry. But when it starts to boil, I have to turn the heat down to avoid splashing. I still want the water to get as hot as possible, though, so it'll cook my stuff fast. Is there any way to make the water hotter without my having to mop up afterwards?

Sorry, but once the water has begun to boil, it's as hot as it will ever get, even if you use a flamethrower. No matter how furiously you might get the water to boil, it will not get any hotter than the boiling point of water: 212°F (100°C), plus or minus.

Let's take a close look at what is happening inside the water as you heat it to boiling.

When you first start heating a pot of water, its temperature rises; that is, the water molecules at the bottom of the pan take on the heat energy and show it by moving faster and faster. After the heat is distributed throughout the water by *convection*, some of the molecules will have so much energy that they can actually break away from their buddies, to whom they had been attached by *hydrogen bonds*. These energetic molecules may even elbow their buddies aside to make bubbles, which then rise and erupt at the surface to release their vapor. We refer to this whole complex process as *boiling*. In summary, the water is taking the heat that you're putting in and using it to change from a liquid to a vapor.

Turning liquid water into gaseous water uses up heat energy, because it takes energy to break the molecules away from each other. If the molecules didn't stick together, water wouldn't be a liquid; it would always be a gas: loose molecules, flying around independently. Every liquid has its own degree of molecule-to-molecule stickiness, and therefore its own breakaway energy, and therefore its own boiling temperature. It requires a temperature of 212°F (100°C) to break water molecules away from each other.

Now let's turn up the heat. The more heat energy per second we pump in from the fire, the more water molecules per second will be acquiring enough energy to break away and shoot off as gas; the water will boil more vigorously and it will boil away faster.

But surprisingly, all that extra heat doesn't make the water's temperature rise, because any extra energy a molecule may acquire beyond

TRY IT

You can check this for yourself by holding a candy or meat thermometer (with tongs) in the steam just above a pot of boiling water and in the water itself, both when it's boiling gently and when it's boiling vigorously. The water temperature will stay the same, but the steam will be a little hotter when the boiling is more vigorous. Moral: You can't cook your spaghetti any faster by turning up the heat. Save your energy.

Bar Bet. No matter how high you turn up the heat under a pot of boiling water, it won't get any hotter.

what it needs to break away, simply goes flying off with the molecule. The extra energy winds up in the steam, not in the liquid that remains behind in the pot. That temperature will remain the same—at the *boiling point*—until all the water has boiled away.

A Water Gate Cover-up

I've noticed that a "covered-up" pot of water boils sooner than an uncovered one. I presume that the pot lid is keeping in some heat that would otherwise be lost, but what kind of heat? There is no hot steam to lose until the water is actually boiling, is there?

Yes, there is. What you're thinking of as steam, the white cloud formed when the water is boiling vigorously, is not steam. *Steam*, also known as water vapor—water in the gaseous form, not solid or liquid—is just as invisible as the nitrogen and oxygen gases in the air. The gases are made of individual, widely separated molecules, flying around independently and randomly. Steam is an invisible gas, or *vapor*. (A vapor and a gas are the same thing; people tend to call a gas a "vapor" if they know that it came directly from a liquid.)

The white cloud, on the other hand, is a *fog*. It is made of tiny droplets of liquid water—not gaseous water—suspended in the air and being carried upward by the rising hot vapor.

There is always some water vapor in the air above liquid water, wherever it is. (You've heard of humidity.) That's because there are always some water molecules at the liquid's surface that happen to be moving vigorously enough to break away from their fellows and fly off. The higher the water's temperature, the more water vapor is produced, because more and more of the molecules will be moving vigorously enough—will be hot enough—to escape. So as the stove heats the water, the number of hot water-vapor molecules in the air above the water increases.

As the temperature of the water goes up, the vapor molecules above its surface are progressively higher- and higher-energy ones, so it becomes more and more important not to lose them. The pot lid closes the gate, so to speak, keeping most of the vapor molecules from escaping and sending them, with their heat energy still intact, right back into the pot. Hence, the water will reach the boiling point faster.

Unless, of course, you watch it.

Too Many Ions in the Fire

I've read that when I add salt to boiling water, it makes it hotter. Sounds impossible to me, but if that really does happen, where does the extra heat come from?

It's strange, but true. The boiling water will indeed begin to boil at a higher temperature as soon as the salt dissolves.

For every ounce of salt that you add to a quart of water (or for every 29 grams of salt per liter of water), the boiling temperature will increase by about 0.9 degree Fahrenheit (0.5 degree Celsius). That's no great shakes, but it's an increase nevertheless. Because the temperature-raising effect is so small, adding salt to your spaghetti water isn't going to cook it noticeably faster. You're adding the salt mostly for flavor, but some people claim that it gives pasta a firmer texture.

TRY IT

With a kitchen thermometer, measure the temperature of a quart of boiling water in a saucepan. It may be lower than 212°F or 100°C because of your altitude above sea level and/or the weather. Now add six ounces (about half a cup) of salt to the water and stir until it is completely dissolved. When the water comes back to a boil, its temperature will be about five degrees (F) higher than before.

The "extra heat" that causes the higher temperature obviously can't be coming from the room-temperature salt that you added. But the burner on the stove is putting out lots more heat than the water really needs to boil. (You can feel that spilled heat all around, can't you?) So there's plenty of heat available for the water to use if it chooses to increase its boiling temperature. The real question is, why doesn't it choose to do so?

A liquid will boil when its molecules get enough energy to break away from each other and go flying off into the air. When salt (sodium chloride) dissolves in water, it splits up into electrically charged sodium and chlorine atoms. (Sodium and chloride *ions.*) These ions do two things:

First of all, they crowd the water molecules, hindering their ability to muscle their way out of the drink and fly off. It's as if the water molecules were trying to get out of a bus by elbowing through a suddenly developed crowd. What they need is an extra shove, an extra amount of energy to help them escape. In other words, they must acquire a higher temperature in order to boil: "to get out of the bus."

The second thing that the charged sodium and chloride ions do is that they gather clusters of water molecules around themselves, which they wear like bulky little wet suits wherever they wander.

Charged particles can attract water molecules because the water molecules themselves are slightly charged: slightly positive at one end and slightly negative at the other; in technical words, water molecules are *polar*, having *two* electric poles: they are *dipoles*. Their positive ends are attracted to the negative chloride ions and their negative ends are attracted to the positive sodium ions. The result of all this clustering is that the sodium and chloride ions have become *hydrated*; they grab onto, and essentially hinder, a large number of water molecules from circulating freely. In order for these hindered water molecules to boil off, they first have to break away from the sodium and chloride ions, and that's much more difficult than simply breaking away from their fellow water molecules, because the *ion-to-H_2O* attraction is much stronger than the *H_2O-to-H_2O* attraction due to *hydrogen bonding*.

Thus, in order to boil off, these hindered water molecules need extra energy. They must achieve a higher temperature. Result: the boiling point has risen.

There's nothing unique about salt, however. Dissolving *anything* in water—sugar, wine, chicken juices, you name it—will produce the same hindrance effect, if not the same clustering effect. So don't ever say that your chicken soup is boiling at 212°F (100°C) just because that's the number you learned in school for pure water at sea level. It's somewhat higher because of all the dissolved stuff in the soup.

Anyway, pure water at sea level boils at 212°F (100 °C) only when the weather is just right.

A Watched Pot Only Simmers

Why do recipes for stews and ragouts *always warn me to simmer them, but not to let them boil? What's the difference, anyway? Isn't a simmer just a slow boil?*

Not exactly. What a simmer aims for is a slightly lower temperature than true boiling, because even a few degrees' difference in cooking temperature can make a big difference in how foods cook. Also, the agitation of a full boil can break up fat-coated, congealed proteins into such tiny bits that they will remain suspended and cloud up your sauce.

In moist cooking—cooking with water present, as opposed to dry roasting—there is only a narrow range of temperatures that you can obtain, so getting just the right temperature can be tricky.

Cooking is, after all, a series of complex chemical reactions, and temperature influences all chemical reactions in two ways: it determines which specific reactions are going to take place, and it determines how fast they will go. Everybody knows the general effect of temperature on cooking speed: the higher the temperature, the faster it cooks. But it's also true that different things happen to food when it is cooked at even slightly different temperatures, because different chemical reactions may be going on.

The question of temperature is particularly important in the moist cooking of meats. Meats undergo different tenderizing, toughening, and drying-out reactions (even when they're immersed in sauces) at different temperatures. The temperature of a full boil, for example, encourages the toughening process, but the slightly lower temperature of a simmer promotes tenderizing. Long experience has taught us which methods work best with which dishes, so it's wise not to monkey with recommended cooking methods.

Outright boiling—lots of big bubbles, such as in cooking pasta—is an infallible indicator of one specific temperature: the boiling point of water. This sets a definite upper limit to the temperature at which we can water-cook food, because water can never get above its boiling temperature, no matter how vigorously we boil it. All those bubbles bursting at the surface are telling us quite clearly that our food is cooking at just about 212°F (100°C), depending slightly on various conditions.

But many desirable cooking reactions take place at lower temperatures. How low? It depends on the food. The only important lower limit for cooking is the temperature that is needed to kill most germs:

about 180°F (82°C). The problem is, how do we achieve one of these lower cooking temperatures reliably when we need it? There is no obvious sign like bubbling to watch for. And we can't be expected to keep sticking thermometers into our pots all the time. When cooks want to tell us that something should be cooked at a certain temperature below actual boiling, they invoke words such as simmer, gentle simmer, slow boil, poach, and coddle. Then they wave their arms around, trying to describe what those words are supposed to mean. And they fail miserably. Look up "simmer" in professional books on cooking technique, and you'll be told that it means everything from 135°F (good luck! Dangerous salmonella bacteria aren't killed until 140°F–150°F), all the way up to 210°F; that is, anywhere from 57°C to 99°C.

But trying to specify a standard "simmering temperature" is futile anyway, because the temperature inside a pot on a stove will vary quite a bit from one spot to another and from one moment to the next. A few of the factors that affect the food's temperature are the size, shape, and thickness of the pot; what it's made of; whether or not it is covered and if so, how tightly; the steadiness of the heat source; the contact between the pot and the burner; the amounts of food and liquid in the pot; and the characteristics of the food itself.

There is only one way to achieve a reproducible and dependable simmer: forget about temperature and concentrate on what the stew is doing. Carefully adjust the pot, the cover, and the burner so that small bubbles are reaching the surface only occasionally. That means that the average temperature in the pot is somewhat below boiling, which is just where you want it. The occasional hot spots here and there throw occasional bubbles to the surface, just to let you know that it's not too cool. Remember that real boiling is when almost all of the bubbles reach the surface. If the temperature is somewhat lower than the normal boiling point, bubbles may form at the bottom, but most of them will be reabsorbed before ascending all the way to the surface. That's a proper simmer.

What about poaching and coddling? Poaching is just another word for simmering, usually applied to fish or eggs. Coddling is placing the food, usually an egg, in water at boiling temperature and then turning off the heat. The temperature decreases steadily as the water cools, so that the average temperature turns out to be the kindest and gentlest of all. The result is a thoroughly pampered, humored, and overindulged egg.

Candy Is Dandy, But More Heat Makes More Sweet

How come sugar syrup gets hotter the longer you cook it, but plain water doesn't?

You've been making candy, haven't you? Obviously, there's something going on in that bubbling sugar-water syrup that's very different from what goes on in plain boiling water.

Candy recipes today tell you to boil the sugar syrup until it reaches certain temperatures on a candy thermometer. Before thermometers became ubiquitous in home kitchens, cooks were told to gauge the temperatures by the behavior of samples taken from the pot and dropped into cold water: does it make a soft or hard ball, a soft or hard crack, etc.? Not only was that a rather subjective test, but different cookbooks gave slightly different temperatures for the various stages.

The boiling temperature of water is raised whenever something, almost anything, is dissolved in it, and sugar is no exception. So any sugar-and-water solution will boil at a higher temperature than plain water will. The more concentrated the solution is, that is, the more dissolved material the water contains, the higher its boiling temperature will be.

For example, a solution of two cups of sugar in one cup of water (yes, it's possible) won't begin boiling until 217°F instead of 212°F (103°C instead of 100°C). But then, as you continue to heat it, many of the water molecules will boil off as vapor and the sugar solution will become more and more concentrated; it's as if more sugar were being added to the water, and its boiling point will continue to rise.

If you boil a sugar syrup long enough, all the water will eventually be gone and you'll be left with nothing but melted sugar in the pot, at about 320°F (160°C). At about the same time, it will begin to *caramelize*: a polite word for the actual breakdown of sugar molecules into a complex assortment of other chemicals that have intriguing flavors in spite of their generally frightful chemical compositions. As *caramelization* proceeds, a yellow-to-brown color transition signals the buildup of more and bigger particles of carbon, which in addition to water vapor, is the ultimate decomposition product of sugar. This transition accelerates, and must be stopped at exactly the right moment. Continue heating just a little too long, and you will be left with a black mass of still sweet, but thoroughly inedible, charcoal.

Eggs Over, Not So Easy

The longer I cook my egg, the harder it gets. The longer I cook my potato, the softer it gets. Why does heat have such different effects on foods?

The short answer is that cooking makes proteins harder and carbohydrates softer. We will leave meats out of this, because the toughness or tenderness of a cut of meat depends in a very complex way on the muscle structure of the animal, the portion of the animal it came from, and on precisely how it's being cooked. During cooking, for example, meat can get tenderer at first and then tougher later on. Your egg vs. potato contrast can be explained entirely by the differing effects of heat on proteins and carbohydrates.

First, let's take a close look at that egg. Eggs are rather unusual in their composition, as befits their unique function in life. If we throw away the shell of a hen's egg and remove the water from what is inside, the dried remains are just about half protein and half fat, with virtually no carbohydrate. The dried yolks are mostly (70 percent) fat while the dried whites are mostly (85 percent) protein. Heat doesn't affect the consistency of fat very much, so we'll concentrate on the protein in the egg white. And you know that we're not going to get out of this without looking at what the molecules are doing, right?

The *albumins* in *albumen* (no typo here; egg whites are called *albumen*, but they contain proteins called *albumins*, with an i) are made up of long, stringy molecules coiled up into globs like very loose balls of yarn. When heated, these balls partially unravel and then stick to each other here and there, making an unholy tangle like a can of spot-welded worms. (Technically, it is said that the molecules are *cross-linked*.) Now when the molecules of a substance change from a bunch of loose balls to an unholy, spot-welded tangle, the stuff is obviously going to lose its fluidity. It's also going to turn opaque, because even light can't get through it.

So liquid egg albumen, when heated above approximately 150°F (65°C), coagulates into a firm, white, opaque gel. The hotter and longer you heat them, the more the molecules will unravel and spot-weld to each other. So the more you cook an egg, the firmer its white will become, ranging from the glop of soft-boiled to the rubber of hard-boiled to the leather of an "over, well" hash-house special. The drying out that takes place at higher temperatures also contributes to the toughening.

The protein in the egg's yolk coagulates in much the same way, but not until it reaches a somewhat higher temperature. Also, its abundant fat acts as a lubricant between the globs of protein, so they can't weld together as much and the yolk doesn't get as tough as the white, even when hard-boiled.

Now about that potato and other foods that contain a lot of carbohydrates. Starches and sugars cook easily; they even dissolve in hot water to speed the process. When you bake a potato, some starches dissolve in the steam.

But there's one very tough and very insoluble carbohydrate that is present in all of our fruits and vegetables: *cellulose*. The cell walls of plants are made of cellulose fibers held together by a cement of *pectin* and other water-soluble carbohydrates. This structure is what makes vegetables such as raw cabbage, carrots, and celery—and potatoes—so firm and crisp. But put the heat on these tough guys, and they turn into wilted wimps. The pectin cement dissolves out into the liquids released by the heat, and the rigid cellulose structure is severely weakened. The result is that cooked vegetables are softer than raw ones.

Meltdown in the Kitchen

Why can I melt sugar, but not salt?

Who says you can't melt salt? Almost any solid will melt if the temperature is high enough. Lava is molten rock, isn't it? If you want to melt salt, all you have to do is turn your oven up to 1474°F (801°C), which will make your kitchen glow a pretty red color. Ovens (steel) won't melt until around 2500°F (1370°C).

Of course, what you mean is that sugar melts much more easily than salt does, that is, at a much lower temperature: sugar will melt at only 365°F (185°C). The question, of course, is why? What's so different about these two common, white, granulated kitchen chemicals? They're both pure chemical compounds and they may look similar, but they're members of two very different chemical dominions.

There are tens of millions of known chemical compounds, each with unique properties. In an effort to make sense out of this vast variety of substances without going completely mad (it has worked for most of them), chemists begin by dividing them into two broad categories: *organic* and *inorganic*. Organic chemicals were originally classified as any substance found in living creatures, and that therefore contained some sort of "life force." But in the early 19th century the "organic" compound *urea*, a component of urine, was obtained from purely inorganic compounds by the German chemist Friedrich

You didn't ask, but . . .
If every pure chemical substance has a specific temperature at which it melts from solid to liquid, does it also have a specific temperature at which it freezes from liquid to solid?

Yes. As a matter of fact, they are identical. The solidification process you describe is what we also call freezing. When we say that water freezes at 32°F (0°C), we could just as well say that that is the melting point of ice. The reason they are the same is that the slithering molecules of a liquid must be slowed down to a certain definite energy in order for them to fall into their permanent, rigid places in a solid crystal. On the other hand, they must be heated up to that same energy in order to break free from those rigid positions and begin to flow as a liquid.

Thus, a certain definite amount of heat is involved in the transition of any substance between its solid and liquid forms. For pure water, that amount of heat happens to be 80 calories per kilogram. If you want to melt a kilogram of ice, you have to put 80 calories of heat into it; if you want to freeze a kilogram of liquid water, you have to take 80 calories of heat out of it.

Just to be contrary, chemists don't call that amount of heat the "heat of melting" or the "heat of freezing." They call it the *heat of fusion*, which I guess is close enough to "melting." To make things worse, whenever a substance happens to be a liquid at room temperature and we have to cool it to make it a solid, people call the transition temperature a *freezing point*. But if the substance is a solid at room temperature and we have to heat it to convert it to a liquid, we call that very same transition temperature a *melting point*.

Wöhler, who wrote to his mentor, "I can make urea without . . . needing to have kidneys, or . . . an animal, be it human or dog." The definition of organic compounds was soon changed to compounds containing the element carbon, which are indeed typical of, but not exclusive to, the compounds found in living things and formerly living things such as petroleum. Inorganic compounds are, well, all compounds that are not organic. Thus, chemicals derived from plants and animals, including sugars, are organic, while minerals, including salt, are inorganic.

If there is a single generalization that can be made about the physical properties of organic and inorganic substances (and, of course, there are exceptions), it is this: organic substances tend to be soft and inorganic substances tend to be hard. The reason for this is that the molecules that organic substances are made of are electrically neutral groupings of atoms, while the components of inorganic compounds are usually *ions*—electrically charged atoms or groupings of atoms. The attraction between opposite electric charges is a stronger force—from two to twenty times stronger—than the attractions of neutral molecules for one another. Thus, inorganic substances are much harder to break apart, to separate their constituent particles from one another, than organic substances are.

Now consider what happens when you heat a solid to its melting point. It is really like breaking the substance apart. The molecules start jiggling around so much from the heat that they eventually begin to flow over and around one another, and that's the definition of a liquid. Obviously, loosely-bound organic molecules should be able to start flowing—to melt—at a lower temperature because they don't require such vigorous agitation to unstick them from one another. So organic substances generally melt at lower temperatures than inorganic substances do.

Sugar (sucrose) is a very typical organic compound, consisting of neutral molecules. Salt (sodium chloride) is a very typical inorganic mineral, made up of sodium ions and chloride ions. It should be no surprise, then, that sugar melts much more easily than salt does.

Nitpicker's Corner. In this book, I use the nutritionist's definition of a *calorie*: the amount of heat energy required to raise the temperature of one *kilogram* of water by one degree Celsius. Chemists have always used a different definition: the amount of energy required to raise the temperature of one *gram* of water by one degree Celsius. The chemist's calorie is thus one thousandth of a nutritionist's calorie; a chemist would call the latter a *kilocalorie*. Sorry about the dichotomy, but it's on its way out. Both tribes now express amounts of energy in the international system (metric, or SI) units of *joules*: one kilocalorie is equal to 4,184 joules. The nutritionist's word *calorie*, the chemist's *kilocalorie*, is so entrenched in the popular mind—that's what's listed in the Nutrition Facts charts on prepared foods—that I use it throughout this book as the most common meaning.

A Toast to Toast!

Why can't I get reproducible results from my toaster? At the same setting, the toast may turn out lighter or darker than the previous time. And different kinds of breads react in very different ways. This is at least the third toaster I've bought in the past several years, and they all work erratically. What's wrong?

What's wrong is that our remarkable twenty-first-century technology still hasn't been able to come up with a toasting method that works consistently. Also, your toasters are lying to you!

This problem has plagued man- and womankind ever since the ancient Egyptians invented bread as we know it and the Romans began browning it over a fire to preserve it. (*Tostus* in Latin means "scorch" or "burn"). Cicero and Lucretius must have debated incessantly in the Forum about who makes the best toast.

It took thousands of years, until around 1909 CE, for electric toasters to make their debut. But they were primitive, clunky, and unreliable. Then, in 1928, Frederick Rohwedder of Des Moines, Iowa, invented a machine that could slice and wrap bread to keep it from going stale. Pre-sliced, pre-packaged bread hit the market in 1930 under the brand name Wonder Bread, which is still one of our most highly esteemed, gourmet-quality breads. (Contrary opinions may exist.)

With pre-sliced bread available, designs and sales of electric toasters took off. But the annoying ritual of scraping the toast usually had to follow its removal from the toaster. Dark brown toast may be tasty, but charcoal is not.

Toast didn't begin to pop up until years later. Today, there are dozens of brands of pop-up toasters on the market, but they all work in much the same way. Pushing the lever down simultaneously switches the electricity on in the heating element (made of a high-resistance metal wire called Nichrome), and lowers a steel "bread elevator" until an electromagnet grabs it and holds it there against a spring's tension. When an electronic circuit (or a bimetallic strip) decides that the job is done, it cuts power to the heater and the electromagnet, whereupon the tray springs up and throws the toast onto the chandelier.

As I stated above, toasters lie. They say you can set the darkness you want, but that's a sham. The machine has no idea of how brown the bread will get; it's color blind. The setting is merely an amount of *time* that the heating elements will stay on before ejecting the bread. It's completely up to you to guess how a given setting on an electric timer—and that's all it is—will produce a given shade of brown. That's why different kinds of bread come out with different degrees of darkness at the same time setting; their original colors, their densities, their moisture contents, and other qualities are all different. And that's also why a second round of toasts of the same kind at the same setting will come out darker than the first: the toaster wires are already hot and don't need any warm-up time.

If toaster manufacturers were honest, they would mark their "darkness" settings in *minutes*. But that may be too much to expect.

So is there no hope? Will technology ever come up with a toaster that can be depended upon to turn out a preset, reproducible color of toast?

You didn't ask, but . . .
What is the connection between toasted bread and a raised-glass, spoken "toast" in honor of a person or event at a banquet?

There are many answers to this question, and most likely, all are wrong.

I can do no better than to quote the *International Handbook on Alcohol and Culture*: "[Toasting] is probably a secular vestige of ancient sacrificial libations in which a sacred liquid was offered to the gods: blood or wine in exchange for a wish, a prayer summarized in the words 'long life!' or 'to your health!'"

True or not, all we can do is honor and enjoy some of the great toasts of history, ignoring the feeble attempts at humor perpetrated at weddings by the besotted "best man."

A simple, heartfelt toast by Robert Burns deserves its fame: "Some hae meat, and canna eat, And some wad eat that want it; But we hae meat, and we can eat, And sae the Lord be thankit."

One of the most verbose toasts ever recorded, at 1,107 words, was "To the Babies" by Mark Twain at a banquet on November 13, 1879, in honor of General Ulysses S. Grant. Mark Twain loved after-dinner performing and took the toastmaster's role as a challenge to his flair for displaying his wit.

On the other hand, one of the humblest and most succinct of all toasts was Jonathan Swift's "*May you live all the days of your life.*"

Perhaps Mr. Clemens could have learned something from Mr. Swift.

Well, maybe. An undergraduate industrial design student at Georgia Tech named Basheer Tome has invented what he calls the Hue toaster, in which a battery of actual color sensors monitors the bread's color and shuts off the heat at the precise shade desired. But as of this writing, it's only a prototype and is nowhere near ready for production. Keep your eyes peeled.

Animal, Vegetable or . . . What?

Nutritionists are always talking about "essential vitamins and minerals." Because the word vitamin comes from the Latin vita, meaning life, it's obvious that vitamins are essential to our health. But what is "essential" about minerals? What are they and why do we need them?

If you have ever played Twenty Questions, you know that everything on Earth has been classified as either animal, vegetable, or mineral. This classification of all of Nature into three "kingdoms" was the brainchild of the Swedish botanist, zoologist, and physician Carolus Linnaeus, who published it in 1735 in his book, *Systema Naturae*.

I need not define animals and vegetables, the two kingdoms of living things. But as a biology-oriented scientist, Dr. Linnaeus cared little for non-living things, so he threw them all into the category he called minerals. He should more properly have called it "everything else," because not all nonliving things are minerals. Are coal, petroleum, amber, glass, pearls, and granite minerals? No. None of them.

What, then, is a mineral? It is a naturally occurring solid substance with a definite chemical composition that can be expressed as a chemical formula. That rules out most rocks, which are conglomerations of bits and pieces of minerals and non-minerals, and do not have specific chemical compositions. Almost 5,000 different mineral species have been identified in the crust of the Earth. Some, we mine for the metals they contain (e.g., hematite for iron, bauxite for aluminum), while others are pretty crystals that we value as gems (e.g., ruby, sapphire, and diamond).

The two most abundant chemical elements in the Earth's crust—the solid outer skin, the miles-thick "egg shell" of our planet—are oxygen, O (46.6 percent) and silicon, Si (27.7 percent). So it's not surprising that about 90 percent of the Earth's crust is made of compounds of silicon and oxygen: minerals with formulas that contain SiO_2 or SiO_4

(*silicate*) groups plus one or more atoms of other chemical elements such as calcium, magnesium, aluminum, sulfur, chlorine, etc.

Some of these elements, whether tied up in silicates or in other chemical compounds, are among the "essential minerals" in our diets. I use quotes around "essential minerals" because it is a misnomer. There are over 200 recognized iron-containing minerals, including hematite, aenigmatite, smaltite, and sleeptite. (No, scratch that last one.) But the chemical element iron itself is the essential nutrient, regardless of the mineral we find it in.

No nutritionist expects us to go out and chew on a rock-hard chunk of aenigmatite (chemical formula $Na_2Fe_5^{2+}TiSi_6O_{20}$—and no, that won't be on the final)—to get our essential dietary iron. All the so-called "essential minerals" should be called *essential elements*, regardless of which minerals or compounds we find them trapped in.

The major elements that we (and all mammals) are made of are oxygen (65%), carbon (18%), hydrogen (10%), and nitrogen (3%), with smaller amounts of calcium (1.5%), phosphorus (1%), and others. You recognize the carbon in organic compounds in plants and animals; the oxygen and hydrogen in water; the nitrogen in amino acids (proteins); and the calcium and phosphorus in bones.

Several other elements are essential for human life. We can divide them into two groups: (1) those that we need in substantial amounts: calcium, chlorine, magnesium, phosphorus, potassium, sodium, and sulfur, and (2) seven to ten others that we need only in trace amounts: chromium, cobalt, copper, iodine, iron, manganese, molybdenum, nickel, selenium, and zinc, although in some cases their roles are unknown and their indispensability is questionable.

But we don't have to go out and chop the minerals out of the ground to get their benefits; the plants and animals that we normally eat already contain them. So if our diets are balanced, we will not suffer from a deficiency of any essential elements.

One widely praised dietary source of essential elements is sea salt. But that is a false claim. The seas and oceans do undoubtedly contain virtually every element you can think of. They have rightfully been called "a weak solution of almost everything." That does *not* mean, however, that these dissolved substances are still present in the edible salt obtained by the solar evaporation of seawater. The methods used to obtain sea salt for human consumption eliminate all elements but sodium and chlorine, plus small amounts of calcium and magnesium.

As the solid sodium chloride crystallizes out of the salt water, whether in a vacuum pan in the case of brine pumped up from an underground salt mine, or in a pool of seawater being evaporated by the sun, the crystallization process itself acts as a very potent purification process. All the other elements, except sodium, chlorine, and small amounts of calcium and magnesium, are left behind in the water when the crystallized salt is filtered or shoveled out.

Thus, while the seas do contain all the essential (and nonessential) elements, those elements simply do not survive the production and purification processes used to separate out the sodium chloride for your kitchen.

Blurp! (Excuse Me!)

I was reheating some pea soup the other day and it started to boil after only a couple of minutes. It was still at a much lower temperature than the boiling point of water. What's going on?

The scientific name (which I just made up) for this phenomenon is *blurping*. Your soup was not boiling; it was merely blurping. Very thick (viscous) semi-liquids can behave quite differently from water and other "thin" liquids.

When you heat a pot of water, the water at the bottom of the pot absorbs heat through its contact with the stove's burner, whether it is gas or electric. (Unless you have one of those high-tech induction ranges, which heat the bottom of the pan magnetically.) Being warmer than the water in the rest of the pot, the molecules in this bottom layer are moving faster—that's what "warmer" means. These faster molecules have to elbow some of their fellow molecules aside to make room for their calisthenics, so the warmed water expands slightly: by 0.0115 percent for every Fahrenheit degree of heating.

The expanded bottom water is now lighter than the upper water and rises through it like a lighter-than-air balloon. Some of the cooler water must then fall down to replace it. These currents of rising and falling water are called *convection currents*. They continue in increased volume, with more hot water rising and more cold water falling to be heated in its turn. This process tends to even out the temperature of all the water in the pot. Eventually, almost all the water in the pot reaches 212°F (100°C), and it boils.

But if the stuff in the pan is thick and viscous, especially as viscous as pea soup or tomato sauce, the rising warm molecules and falling cool molecules cannot move easily through the muck. (Nothing personal about your pea soup.) So convection is impeded, and the hot muck stays pretty much down at the bottom, getting hotter and hotter until a part of it erupts into a big steam bubble that blurps and splatters all over you and the kitchen walls and ceiling. (I exaggerate.) But that is not boiling; the soup in the higher regions of the pot may still be refrigerator-cold when the blurping begins at the bottom. How can you keep yourself and your kitchen unsouped? You must provide a substitute for the normal convection currents. With your own elbow grease, you create artificial convection to keep the temperature uniform throughout the pot. In other words, stir, stir, and stir while heating.

Noisy Water

This may be a stupid question, but when I heat a pot of water on the stove, why does it make sizzling and squealing noises before it comes to a boil?

There is no such thing as a stupid question; there are only stupid answers. Hoping not to add to the world's all-too-abundant supply of the latter, I hereby present my analysis of this less-than-earthshaking phenomenon.

The heating of a pot of water is far from uniform. Convection currents keep the water at a uniform temperature all the way up to the boiling point, but there will always be some extra-hot spots down at the bottom of the pan, where the burner is feeding in heat and convection cannot keep up with it. Long before the water has reached the boiling temperature, some of the hottest spots at the bottom of the pan will burst out into invisibly small bubbles of vapor that don't have enough oomph (buoyancy) to rise, so they simply collapse *in situ*.

The constant production and reabsorption of these tiny bubbles tap out a miniature drumbeat, a sizzle, as the bubbles pop in and out of existence like a chorus of tiny, exploding balloons. As the steam in a bubble condenses and liquid water rushes in to fill the void, vibrations are created that propagate as sound. Because the bubbles are so small, the sound is a high-pitched sizzle or squeal, rather than the lower-pitched glub-glub that will come later, when bigger bubbles burst at the surface and boiling triumphs.

The Rice Must Be Precise

I want to make a paella, but I'm told that it requires a special kind of rice. Recipes I've seen specify a short-grain rice, but which one? And how can I tell if a rice is short-grain or long-grain?

Get a ruler. I'm just kidding, but not entirely. The length of the grain really does mirror the cooking characteristics of any given type of rice. Because there are more than 40,000 known varieties of rice, trying to name them all would be maddening, so we just refer to them as short grain, medium grain, and long grain. It works quite well.

Rice grains are the seeds of two plant species: *Oryza sativa* (Asian rice) or *Oryza glaberrima* (African rice). Rice feeds nearly half the world's population, and about 90 percent of it—more than 400 million tons per year—is produced and consumed in Asia.

But rice plays a major role in the cuisines of many other parts of the world, including the Caribbean and the Arab countries. In fact, it was the Moors in the 10th century who introduced rice to Spain, where it became established in Valencia, which is still the center of rice farming in that country.

Paella is all about the rice, not the hodgepodge of chicken, sausage, or seafood that restaurants may stick into it. The three main varieties of paella-worthy Spanish rice are Bahia, Senia, and Bomba. In the center of each of these grains is a *perla* (pearl), a dense nucleus of concentrated starch that imparts a certain toothiness to the cooked grain. But of all the varieties, only two, Bomba and Calasparra (also called Sollana), are celebrated for making the finest paella. They are both grown near the town of Calasparra in the region of Murcia, near Valencia. Bomba is by far preferred, but is expensive even in Spain and not easy to find in the US.

It may well be said (and I shall say it) that *the shame in Spain comes mainly from the grain*: the wrong kind of grain used by uninformed cooks. Paella rice must be highly absorbent to soak up all the savory seafood or poultry juices as it cooks. It must also be just sticky enough to be spooned out of the pan and deposited in loose clumps on the plate. Bomba excels at both these qualities. Chopstick-friendly Asian rice, bred to be very sticky, will not do, nor will the American preference for separate, fluffy long grains.

And by the way, what is called Spanish rice in American Mexican restaurants is a mélange of any-old rice, onion, tomato sauce, bell pepper, and chili powder. It is mercifully unknown in Spain.

The various varieties of rice are classified in three categories according to size: short-grain (less than 5 millimeters long), medium-grain (5 to 6 millimeters long), or long-grain (6 to 7 millimeters long). While this classification seems to be evading more fundamental distinctions, it does coincide roughly with the types of starch they contain, and therefore with their behavior in cooking.

There are two major kinds of starch in rice: *amylose* and *amylopectin*. Most rice contains a mixture of the two. Amylose molecules are long, stringy chains, while amylopectin molecules are shaped like branches of a tree. Chinese and Japanese short-grained rice is rich in the branched molecules of amylopectin starch, which can trap lots of water; that makes the rice sticky when cooked. American long-grain rice, on the other hand, is richer in the straight-chain amylose starches that cannot trap much water, so the cooked grains don't stick to one another as much. Italian Arborio rice is bred to be creamy for risotto, while Asian rice is meant to be sticky. The medium-grain rice used for sushi has just enough stickiness to make it moldable into a solid shape.

Paella? The ideal paella rice must absorb lots of the highly flavorful broth, yet become only slightly sticky in the process. Bomba fits the bill perfectly. It is a short-grain rice that can absorb three times its volume of broth when cooked in the paella pan. Other short-grain rices top out at about twice their volume, while long-grain rices can absorb only their own volumes. That's why when we cook long-grain rice (the common American kind) we add only an equal volume of water, and for most short-grain types, we add twice as much water. When making authentic paella with Bomba or Calasparra rice, we add three times the volume of water and stand there as it simmers, worrying about whether the finished product will be too soupy. It will not; all that water will be absorbed. Bomba has the peculiar property of expanding in width like an accordion, instead of in length.

Can't find or afford Bomba? Senia, Bahia, and Calasparra rice are just about as good and Arborio can substitute in a pinch. But no, you can't use long-grain or Asian varieties. And please spring for genuine saffron as the spice and colorant, not the turmeric or (horrors!) food coloring used in most American restaurants.

¡Buen apetito!

Those Naughty Microwaves

In a magazine article I read that cooking broccoli in a microwave oven destroys its nutrients. Is that true?

This canard has been circulating ever since 2003, when a group of researchers in Spain published a study that seemed to show that microwaves kill the nutrients in broccoli and presumably in other vegetables as well.

(A little etymology is interesting here. You undoubtedly know that a canard is a groundless rumor or belief, while "canard" is the French word for duck. But how did those two meanings become related? Look it up, as I did, and you will be offered a choice of several different theories, all of them no better than guesses. My favorite one is that a French prankster wrote a newspaper story in 1896 about cannibal ducks eating one another alive. The story "went viral" as we would say today, and American newspaper editors began calling any hoax story "a canard.")

But back to broccoli.

Very soon after the publication of this research, a London-datelined news story trumpeted, "Microwave blasts out broccoli's health benefits." A subsequent article in *Prevention* magazine was headlined "Nuking Broccoli a No-No." And dozens of other media stories carried the study's purported finding that microwaving destroys broccoli's nutrients.

I obtained a copy of the original research report, a document that would quickly put to sleep anyone uninitiated in nutritional chemistry. Without putting my readers to sleep, I can simply state that the research proved no such thing. It was a poorly designed study, reported rather ambiguously. What the researchers did find was that the more water the broccoli was cooked in and the longer it was cooked by any method, the more nutrients—specifically the antioxidant flavonoids and flavonoid derivatives—were extracted into the water. That's no surprise, because flavonoids, as well as vitamins C and the B vitamins, are soluble in water.

The moral, then, is that we should cook our vegetables for as short a time and with as little water as we can get away with. We can't do without heat in cooking, but we can reduce the amount of water.

Well then, how about using no water at all? After all, raw broccoli is already 89 percent water.

I have experimented with the completely water-free microwaving of broccoli, cauliflower, and Brussels sprouts, with excellent results. Here is the procedure: Rinse the vegetables in cold water and dry with paper towels or a clean kitchen towel. Place the vegetables in a small, covered, microwave-safe container. Do not add water. Zap at full power for three to four minutes or until the vegetables are tender. Do not overcook.

And that's it. The vegetables will have cooked in their own steam, with no loss of nutrients except, perhaps, as caused by the heat. Vitamins C and the B group are not only soluble in water, but are unstable when heated by any method, whether by boiling, steaming, or microwaving.

How, then, can I explain the Spanish researchers' purported finding that microwaving broccoli destroyed almost all of its flavonoids? I can't, and neither can they. As the authors themselves confess, "This flavonoid loss rate does not agree with that previously reported by other authors for microwaving."

All of which illustrates an even more important moral: Neither journalists nor health-conscious consumers should go running around like Chicken Little because of a single research study—even if it had been carefully designed and accurately reported in a peer-reviewed journal. We must remember that every "latest study" will someday be replaced by a later "latest study" that might either confirm it or refute it.

That's how science progresses; it zigs and zags towards the truth.

On Puffery and Shreddery—How Do They Make Puffed Rice?

Quaker Puffed Rice has been around since 1904, when it was introduced at the St. Louis World's Fair. But many other grains such as wheat, barley, and buckwheat can be made to explode into expanded, puffed-up forms. Why do all these things "Snap, Crackle, and Pop"?

Kellogg's Rice Krispies, another venerable puffed rice cereal, has been snapping, crackling, and popping on the market since 1928.

And you have heard, perhaps, of popcorn? Popcorn consists of whole kernels of corn dried to an ideal moisture content of 13.5

percent. When heated in oil or air or by microwaves, the water vaporizes and builds up steam pressure. When the pressure gets to about 135 pounds per square inch, it bursts the kernels' tough hulls and steam shoots out along with the heat-gelatinized starch, which instantly cools and solidifies into that splattered shape we all love.

Puffed Rice isn't made in the same way as puffed, or popped, corn. Rice grains don't contain any water to speak of, and they don't have the tough, airtight hulls that corn kernels do, so they can't be popped by corn's steam-bomb method.

You didn't ask, but . . .
How do they make Cheerios into those perfect little donut shapes?

Cheerios are more complex than simple puffed grains. They are made from cooked oats, corn and wheat starch, sugar, and salt, plus a bunch of vitamins and minerals and stuff (read the ingredients on the package), all mixed with water to make a dough. As in making macaroni, the dough is extruded through a die to make a tube, which is instantly chopped into tiny donuts by rapidly rotating blades. After partial drying, the donuts are puffed by a pressure-vacuum gun system similar to Quaker's Puffed Oats. After more drying and a little toasting, they're ready to pour into your breakfast bowl.

Okay now, what about the most peculiar cereal shape of all: Shredded Wheat?

In 1892, an Ohio lawyer named Henry Perky invented a machine to make what he called "little whole wheat mattresses," in the hope that they would stem his troublesome diarrhea. History does not record whether they worked for him, although for more than 120 years millions of others have been enjoying them for less urgent purposes.

Boiled and cooled wheat is rolled between two special metal rollers: one smooth and the other containing many grooves around its circumference. As the rollers roll, a toothed comb scrapes the mixture out of the grooves, leaving a sheet of thin ribbons of wheat paste. The sheets fall onto a moving belt that lays them on top of one another in stacks. The stacks are then cut into a variety of pillow shapes—loaves, minis, etc.—and baked to make them toasty and crunchy.

Please note that I have refrained from making a joke about Mr. Perky's name or that of his brother, Herky.

At Kellogg's, they expose the rice grains to steam to infuse the necessary water into them. Then, they heat them in an oven to convert the water into steam, which escapes irregularly, leaving sponge-like holes, caves, and tunnels. Cereal scientists have figured out that when you pour milk on Rice Krispies, the milk gets absorbed into these caves and forces the air to find a way out. If its path is blocked, the air must break its way out through the caves' walls. Thus, the snaps, crackles, and pops that we hear. Except in Germany, of course, where Rice Krispies say "*Knisper, Knasper, and Knusper.*"

The Quaker Oats Company uses a different grain-popping method to make its Puffed Rice and Puffed Wheat. In an airtight enclosure, they infuse the grains with moisture, using steam or hot water. Then, they increase the air pressure inside the enclosure to around 200 pounds per square inch and release the pressure suddenly, as if puncturing a balloon. The water inside the grains instantly vaporizes and blasts out like the gases that propel a bullet through a gun. The grains are left full of holes. They've been puffed.

CHAPTER 3

Around the House

Let's just wander about the house for a while. If we have our antennas up, we will find a wealth of fascinating things to look into. Perhaps candles are burning on the dining table and the champagne is bubbling away while you admire a sunset through the picture window. Or maybe you're stuck down in the laundry, where soap and bleach are working their chemical magic on all that miscellaneous stuff that we lump under the general heading of "dirt."

In this chapter, we will see some astounding things that are going on in the candles, the champagne, the sunset, the soap, and the bleach, not to mention your water bed and your shower.

Burning a Candle at One End

When a candle burns, where does the wax go?

Except for what drips all over your tablecloth, it goes to the same place that gasoline and oil go when they burn: into the air. But in a chemically altered form.

Candles are usually made of paraffin, which is a mixture of hydrocarbons, organic compounds that we find in petroleum. As the name implies, hydrocarbons contain nothing but hydrogen atoms and carbon

Nitpicker's Corner. When there isn't quite enough oxygen available to make a full complement of carbon dioxide, as in an automobile engine, we get some carbon monoxide also.

atoms. When they burn, they react with oxygen in the air. The carbon and oxygen become carbon dioxide, while the hydrogen and oxygen become water. (But not necessarily completely.) Both of these products are gases at the temperature of the flame, and they just go up into the air.

We burn many other hydrocarbons: methane in natural gas, propane in gas grills and torches, butane in cigarette lighters, kerosene in lamps, and gasoline in cars. They all burn to form carbon dioxide and water vapor, seeming to disappear in the process. Paper, wood, and coal contain additional mineral plant materials that do not burn, so besides producing carbon dioxide and water they leave an ash.

TRY IT

If you find it hard to believe that flames produce water, try this: Put some ice cubes in a small, thin aluminum saucepan, let it get cold, and hold it just over the flame of a candle or a butane cigarette lighter. After a while, check the bottom of the pan and you will see that water vapor from the flame has condensed there into liquid water.

You didn't ask, but . . .
Why won't a candle burn without a wick?

By capillary attraction, the wick leads melted wax up to where it can be vaporized to mix with oxygen in the air. A block of solid wax, or even a puddle of melted wax, will not burn because the wax molecules cannot come into contact with enough oxygen molecules; only as vapors can they mix intimately, molecule for molecule, and react. Combustion (burning) is a reaction that releases heat energy. Once it begins, it gives off more than enough heat to keep melting and vaporizing more wax to keep the process going.

Fire!

The flames in my gas grill are blue, but the candles on the dinner table burn with a yellow flame. What makes flames different colors?

It's a matter of how much oxygen is available to the burning fuel. Lots of oxygen makes blue flames, while limited amounts of oxygen make yellow ones. Let's look at the yellow flame first.

A candle is really a very complex flame-producing machine. First, some of the wax must melt, then the liquid wax must be carried up the wick, then it must be vaporized to a gas, and only then can it burn—react with the oxygen in the air to form carbon dioxide and water vapor. This is far from an efficient process.

If the burning were a hundred percent efficient, the wax would be transformed completely into invisible carbon dioxide and water. But the flame cannot get all the oxygen it needs to do that, just by taking it out of the air in its immediate vicinity. The air, with its flame-nourishing cargo of oxygen, just cannot flow in fast enough to take care of all the melted and vaporized paraffin that is ready to burn. So under the influence of the heat, some of the unburnable paraffin breaks down into tiny particles of carbon, among other things. These particles are heated by the flame and become luminous; they glow with a bright yellow light. And that's what makes the flame yellow. By the time the glowing carbon particles reach the top of the flame, almost all of them have found enough oxygen to burn themselves out.

The same thing happens in kerosene lamps, paper fires, campfires, forest fires, and house fires: yellow flames, all. Air just cannot flow in fast enough to make the fuels burn completely to carbon dioxide and water.

TRY IT

If you don't believe that there are tiny particles of unburned carbon in a candle flame, just insert the blade of a table knife in the flame for a few seconds, to catch them before they burn out. The blade will acquire a deep, velvety black coating of carbon. This carbon black is just about the blackest substance known, and is used in inks.

Gas grills and gas ranges, on the other hand, start out with a gaseous fuel. No vaporizing required. That makes it easy for the fuel to mix with lots of air, so that the burning reaction can go full blast. Because all of the fuel is burning completely, we get a much hotter flame. And it is a clear, transparent flame because there are no glowing carbon particles cluttering it up.

Want hotter yet? Why not mix pure oxygen, instead of air, with the fuel gas? After all, air is only about 20 percent oxygen. Glassblowers use a torch that mixes oxygen with natural gas (methane), to produce a flame temperature of about 3,000°F (1,650°C). A welder's oxyacetylene

You didn't ask, but . . .
Why should a hot, well-adjusted gas flame be blue, instead of some other color?

It has to do with the fact that atoms and molecules that are heated in flames can absorb some of the heat energy and then promptly spit it back out as light energy.

Every substance has its own typical wavelengths or colors of light that it emits after being stimulated by heat. (Every substance has its own unique emission spectrum.) The propane or natural gas in your gas grill and the acetylene in the welder's torch are very similar; they are all hydrocarbons—compounds of carbon and hydrogen. It happens that hydrocarbon molecules emit many of their particular light wavelengths in the blue and green parts of the visible spectrum. Other kinds of atoms and molecules, if they were vaporized and burned, would impart their own particular colors to the flame. That's how colored fireworks are made.

(oxygen plus acetylene gas) torch can reach about 6,000°F (3,300°C). Clear, blue flames, all, except when the torch is misadjusted, so that the gas doesn't get enough oxygen to burn completely. Then? A yellow, sooty flame.

Here's the Dirt on Soap

They say there are three things you don't want to see being made: sausage, laws, and soap. I have already heard too much about (and from) legislators, and I would rather not know about sausage. But I will brace myself: how do they make soap?

The unholy mess involved in making soap belies its use over at least the past 2,000 years as an incomparable cleaner of just about everything. It has always been easy to make out of cheap materials: fat and wood ashes. Lime was sometimes used also.

Here's how the Romans did it: Heat limestone to make lime. Sprinkle wet lime onto hot wood ashes and mix well. Shovel the resulting gray sludge into a cauldron of hot water and boil it up with chunks of goat fat for several hours. When a thick layer of dirty brown curd forms on the surface and hardens upon cooling, cut it into cakes. That's your soap.

Or if you prefer, just go to the store and buy a cake of today's highly purified commercial product. In addition to soap, which is a definite chemical compound, it probably contains fillers, dyes, perfumes, deodorants, antibacterial agents, various creams and lotions, and lots of advertising. Sometimes more advertising than soap.

Every soap is made by the reaction of a fat with an alkali—a strong, potent base. (A base is the opposite of an acid.) Instead of goat fat, today's soaps are made from any of a number of different fats, including beef or lamb tallow and the oils of palm, cottonseed, or olive. Castile soap is made from olive oil. The alkali is usually lye (caustic soda, or sodium hydroxide). Lime is another handy alkali, while wood ashes can still be used in a pinch because they contain the alkali potassium carbonate.

Having been created by the addition of an organic compound (a fatty acid) to an inorganic compound (lye), the soap molecule retains some features of both its parents. It has an organic end that likes to fraternize with oily organic substances, and an inorganic end that is attracted to water. Hence its incomparable ability to coax oily dirt into the wash water.

Whenever you see the following chemicals listed as ingredients on the label of a shampoo, toothpaste, shaving cream, or cosmetic, be neither alarmed nor impressed; they are all just the chemical names of soaps: sodium stearate, sodium oleate, sodium palmitate, sodium myristate, sodium laurate, sodium tallowate, and sodium cocoate. If the "sodium" is replaced by "potassium," the soap has been made with caustic potash (potassium hydroxide) instead of with caustic soda (lye or sodium hydroxide). Potassium soaps are softer, and may even be liquids.

Cleanliness Is Next to . . . Impossible

Whenever there is something on our bodies, clothes, or cars that we do not like, we say they are "dirty" and we wash them. What we call "dirt" can be any kind of stuff at all. But soap always seems to oblige us by removing it, and only it. How does the soap know what's dirt?

It would appear that soap recognizes and respects our skins and precious possessions, while devouring everything else under the sun like vultures leaving only bones behind. But no such magic substance exists. Instead, the answer has to do with the natures of oil and water. Simplistic as it may sound, everything that we call dirt—more politely, "foreign substance"—is either oily or is stuck to us with oil. And soap is a uniquely good oil remover.

Before we can figure out how to remove dirt, we must look at how we get dirty in the first place.

A microscopic speck of dirt, meaning anything we do not want stuck to us, can be stuck to us in one of two ways: it is mechanically trapped in a microscopic crevice or else it is moist, and the moisture makes it adhere. An example of the former is the kind of dirty you get on a dusty road; an example of the latter is what can happen on a muddy one. In either case, a good hosing down with plain water, encouraged perhaps by a little rubbing, will do a pretty good job of removing the foreign substances. Soap is not really necessary.

Nitpicker's Corner. There is a second important thing that soap does: it makes water wetter. That is, it helps the water to penetrate into all the nooks and crannies of whatever it is we're washing.

Water molecules stick to each other quite strongly. One result of this fact is that a water molecule that is situated right at the surface of a "piece" of water has very strong attractions that are trying to pull it into the rest of the "piece." Now the tightest huddling-together formation that any group of particles can achieve is to gather into a spherical shape; a sphere has the smallest possible amount of surface exposed to the outside world. That's why water forms spherical drops whenever it is free to do so, such as when it is falling as rain.

(In two dimensions, that's why the pioneers "circled the wagons" against attacking Indians; if they had "squared the wagons," they would have been exposing more of themselves to the outside.)

This inward-pulling force on the surface molecules of a liquid is called surface tension. It arises because the surface molecules are, in a way, different from the molecules in the body of the liquid.

In the body of a liquid, a molecule is pulled upon by attractions to fellow molecules above, below, and all around it, and all of these pulls cancel each other out. It's like a tug of war pulling in all directions: nobody wins. But a molecule right on the surface is pulled upon only from below, and all around, but not from above; so there is a net downward pull, not cancelled by any upward pull. This makes the surface molecules adhere more tightly to the bulk of the water than the other molecules do, and the water behaves as if it had a taut skin on its surface. Small objects can even lie on the surface without sinking through the "skin." Water bugs can even skate merrily along on it.

But what if the dirt particles have a slightly oily coating instead of a watery one? They will stick to your skin just as the wet mud did. In fact, the dirt does not even have to bring along its own oily coating; there is often enough oil on our skins to make dirt particles stick. Unlike the mud, however, this dirt is going to stay stuck, because oil doesn't evaporate and dry up, as water does. Nor will a spray of plain water dislodge it, because water will not have anything to do with oil; it will simply roll off the dirt as if off a duck's back, which, as you know, is covered with oily feathers.

It seems, then, that the only thing we can do to unstick oil-adhering dirt is to seek and destroy the sticky oil itself. The dirt will then be able to fall off or be swept away by a liquid.

Well, then, let us fill the old bathtub with alcohol, kerosene, or gasoline; they are all good solvents for oil, aren't they? That's what dry cleaners do to our dirty clothes: they tumble them around in a barrel full of a solvent such as *perchlorethylene*, or *perc* for short, an organic solvent that is a phenomenal solvent for oil. (They call the process "dry" cleaning in spite of the fact that it involves sloshing the clothes around in a liquid. The thinking seems to be that if it isn't water, it isn't wet. Wrong, of course.)

Unfortunately, perc in the bathtub would kill you even faster than the alcohol, kerosene, or gasoline would, so we can forget about bathing in solvents. But there is one substance that is just as good, and it's not very toxic (mouths have reportedly been washed out with it): soap. Soap doesn't actually dissolve oil. It accomplishes the astounding feat of enticing the oil into the water, so that it and its captive specks of dirt can then be flushed away.

Soap molecules are long and stringy. For almost all of their length (the "tail") they are very similar to oil molecules, and therefore they have an affinity for other oil molecules. But at one end (the "head") they have a pair of ions, which just love to associate with water molecules, and this head is what drags the whole soap molecule into the water—makes it dissolve.

While swimming around in the water, if a gang of dissolved soap molecules encounters an oily particle of dirt, their oil-loving tails will latch onto the oil, while their water-loving heads are still firmly anchored in the water. The result is that the oil is pulled into the water; its captive dirt particle is released from whatever it was stuck to and can be swept down the drain.

TRY IT   You can rest a steel sewing needle on the surface of water in a bowl. Lower it down carefully with a couple of toothpicks or match sticks.

Enter soap. Soap molecules disrupt the surface tension of water by crowding around the water surface, with their water-loving heads in the water and their oil-loving tails sticking out. This disrupts the huddling-together tendency of the water molecules and allows them to pay some attention to—that is, to adhere to and wet—other things.

Including a floating needle.

After you've gotten the needle to rest on the water's surface, sprinkle some powdered laundry detergent near it, but do not actually bomb it. Detergents are even better than soap at killing surface tension. As soon as some of the detergent dissolves, the needle will plunge precipitously to Davy Jones' locker.

Love-Boat Logic

A TV commercial for a Caribbean cruise line says, "For our guests who may have been out in the sun too long, we even wash our sheets in soft water." Is that for real?

No, the advertising copywriter has been out in the sun too long. Makes you wonder whether a cruise line that swallows this kind of foolishness from its advertising agency is capable of finding the right islands.

Rather than insulting my readers by pointing out why the sheets will not be any softer, I might gently remind any prospective cruisers among them that hard and soft water are not called that because of their relative rigidities. Nor is it because you make hard-boiled eggs with one and soft-boiled eggs with the other. The choice of "hard" and "soft" as sobriquets for these waters was unfortunate; they could better have been called "difficult" and "cooperative"—with respect to soap.

Hard water is water that has been around the block. It first fell through the air as rain, and then frolicked and percolated over, around, and through the rocks and rills before being apprehended, detained, and exploited by humans. In its peregrinations, it inevitably picked up carbon dioxide from the air, which made it acid: carbonic acid. This acid can dissolve small amounts of calcium- and magnesium-bearing rocks such as limestone (calcium carbonate) and dolomite (mixed calcium and magnesium carbonates). Certain iron-bearing minerals can also dissolve slightly. As a result, the water can wind up containing dissolved minerals such as calcium, magnesium, and iron.

What's hard about it is that it is hard for soap to do its job properly in water that contains these minerals. Soap consists of long molecules with an oil-loving end and a water-loving end. It does its cleaning job primarily by joining oil and water together. The trouble is that calcium, magnesium, and iron react with the water-loving ends of the molecules to form insoluble, white waxy curds that effectively remove the soap from the water and prevent it from doing its job. The curds are evident as "soap scum" or, in its most infamous guise, the "bathtub ring". (Contrary to folklore, the latter is more a sign of the hardness of the water than of the hygienic habits of the bather.)

Incidentally, here is our candidate for the most disquieting thought of the week: you have probably eaten soap scum in certain candies. One common form of soap scum goes by the chemical name of *magnesium stearate*; the stearate part comes from the soap, while the magnesium part comes from the hard water. Magnesium stearate is a soft, smooth, waxy substance. That's what makes it stick to the bathtub, all right, but that's also why it imparts a creamy texture to soft mints and other "suckable" candies—a soapy texture, if we may be so bold. But if you see magnesium stearate listed among the ingredients of your candy, be assured that it is the pure chemical compound, manufactured from sources quite different from bathtub scrapings.

But back to the hard water. We can do two things about soap's inability to do its job well in hard water: we can soften the water or we can lose the soap and use a synthetic detergent instead.

Water-softening tactics are based upon removing the offending minerals or by rendering them ineffective. Many home water-softening units remove the minerals by ion exchange. Ion exchangers replace the calcium, et al. with sodium, which is innocuous because it is already a part of the soap molecule.

In what seems like ancient times, about 50 years ago, hard water was combatted by adding *washing soda* (sodium carbonate) to the laundry tub. This chemical re-forms the original, insoluble calcium and magnesium carbonates—essentially, the original rock—thereby removing them before they can gum up the soap. These days, however, practically nobody uses soap for laundry. The acres of wash-day products on the supermarket shelves are all synthetic *detergents* (and are all essentially identical, except for the hype). Like soap, they have oil-loving and water-loving ends on their molecules, but they simply refuse to react with calcium or magnesium. For good measure, they

TRY IT

Shake up a few shavings of bar soap or a few flakes of Ivory Snow (which is real soap) with some distilled water in a jar. You'll get a beautiful, thick head of suds, indicating that the soap is doing its job. (Distilled water is pure and free of minerals; you can find it in many supermarkets and drug stores.)

Next, if you live in a hard-water area, add some tap water and shake again. (If your water supply is soft, you can simulate hard water by adding a little milk instead.) The calcium in the hard water (or milk) will kill the suds flat. You may even see some soap scum in the form of floating white curds.

usually contain water softening chemicals such as phosphates and—guess what?—washing soda.

Hard water is still a villain, however, because it can clog up water pipes and boilers. When hard water is boiled, the dissolved calcium and magnesium fall back out of the water as limestone and dolomite. This born-again rock—called boiler scale—can form a tenacious coating on the insides of boilers, water heaters, and pipes, clogging them up like the arteries of a Viennese pastry chef. If your water supply is hard, shine a flashlight inside your dry tea kettle and you'll see the boiler scale as a white coating on the surface. If it bothers you, boil some vinegar—an acidic liquid—in the kettle to dissolve it.

A Whoosh Is No Big Whoop

I usually buy my soda pop in two-liter bottles. But with such a big bottle, keeping the leftovers alive and fizzy from one pizza to the next is a problem. Besides keeping the cap on, what else can I do to keep it from going flat? What about that gadget that you screw onto the bottle and pump up? Does it really work?

Your objective is to keep as much carbon dioxide gas in the bottle as possible, because that's what the fizzy bubbles are made of. Keeping the bottle tightly stoppered certainly has to be your first line of defense. But frankly, it won't help very much.

There are many kinds of stoppers on the market, including that fancy pump-up job that you mention. It is essentially a miniature bicycle pump that you screw onto the bottle, and then you pump a plunger to compress the gas inside the bottle. Sounds good, but unfortunately, it's a complete fraud. All it does is make you think your soda is livelier than it is. Let's see why.

Soda fizzes when dissolved carbon dioxide gas emerges as bubbles. The gas wants desperately to escape from the liquid because the folks down at the bottling plant have pumped in much more carbon dioxide than would ordinarily dissolve under atmospheric conditions. As soon as you open the bottle, most of that excess gas escapes into the room, and there is absolutely nothing you can do about that. Your only problem is how to make the remaining gas stay in the liquid for as long as possible.

Three things determine how much of a gas can remain dissolved in a liquid: the chemical reactions of the specific gas, the pressure, and the temperature.

• Reactions: Gases that react chemically with water will generally dissolve more readily than inactive gases, whose molecules have nothing to do but cruise aimlessly around in the water. But carbon dioxide is one of those gases that react; it forms carbonic acid, which adds that nice little pungent taste to soda, beer, and sparkling wine. Air (nitrogen and oxygen) doesn't react with water. As a result, carbon dioxide at room temperature is more than 50 times more soluble in water than nitrogen is, and more than 25 times more soluble than oxygen.

• Pressure: The effect of pressure is just what you would expect: the higher the gas pressure above the liquid, the more gas will be pressed into the liquid. The way it works is that at higher pressures there are more gas molecules flitting about per cubic inch in the space above the liquid, and more of them will therefore be diving each second into the liquid.

• Temperature: The effect of temperature is probably just the opposite of what you would expect: the higher the temperature, the *less* gas will dissolve. Saying it the other way, the colder a liquid is, the *more* gas it can hold. The reason for this is a little more involved than we want to get into right now, so we'll save it for later. But one example: at room temperature, water can hold only about half as much dissolved carbon dioxide as it can at refrigerator temperature.

Our conclusions, then, are that in order to keep as much carbon dioxide dissolved in the soda as possible, we must keep the gas pressure high and the temperature low. Temperature is no problem; we will just make sure it is good and cold before we open the bottle, and then we will put the leftovers back in the refrigerator as soon as possible.

But pressure is quite another matter. At the bottling plant, the carbon dioxide molecules were forced into the soda like a crowd of claustrophobes into an elevator. The instant we open the bottle, a frantic exodus takes place, and virtually all the carbon dioxide pressure goes off in one big whoosh! Once that happens, your soda is inevitably going to flatten; it is just a matter of time.

But is there really nothing we can do about that? Can't we restore the pressure somehow, that we may live to belch another day?

Enter the gadget hucksters. Just screw their gizmo onto the bottle, they say, pump the piston a few times, and there you are. Next time you open the bottle, you'll be treated to the biggest, most satisfying whoosh! you ever heard, and you are supposed to think that your soda is factory fresh.

You didn't ask, but . . .
Why does warm beer go flat?

More gas—any gas—can dissolve in a liquid when it is cold than when it is hot. Or as a chemist would say, the solubility of a gas in a liquid increases with decreasing temperature. In practical terms, why does the carbon dioxide choose to leave the beer just because it is getting warmer?

From everyday experience, you might expect that as a liquid gets warmer, it should be capable of dissolving more stuff, not less. You can dissolve more sugar in hot tea than in iced tea, can't you? Then why should gases be any different?

The answer lies in the role that heat plays in the dissolving process. It can be a very complicated one.

When a substance dissolves in water, its molecules separate from one another and disperse themselves throughout the water. Other changes may occur at the same time, depending on the substance that is dissolving. For example, the molecules might attach themselves to tight little clusters of water molecules, or they may react chemically with the water, or they may split into electrically charged fragments (*ions*) or do other things too horrible to contemplate. (Metallic sodium can explode when thrown into water.)

All of these processes either use up or give off energy in the form of heat. So heat plays an intimate and widely varying role in the dissolving of various substances. The net result is that some substances will eagerly absorb the extra heat in hot water and use it to dissolve more, while some substances will react negatively to the extra heat and dissolve less. In other words, some substances will be more soluble in hot water than in cold, and some will be less soluble in hot water than in cold. Even chemists can't always predict which way it will go for a given substance.

In the case of gases, though, we can generalize: when gases dissolve in water, they all give off energy in the form of heat. So they will dissolve more readily in a cold, heat-absorbing environment like cold water, and they are discouraged from dissolving in a hot, heat-rich environment like hot water.

But guess what? There is not one more molecule of carbon dioxide in there than if you had simply screwed the cap on tight. You would get the same big whoosh! if there were nothing but plain water and air in the bottle. The gizmo is nothing but an expensive stopper.

What you've pumped into the bottle is air, not carbon dioxide. Sure, there's a little carbon dioxide in the air, but it's only about one out of every three thousand molecules. The escape of a gas from a liquid can be decreased only by putting more of that particular gas into the space above the liquid. The amount of carbon dioxide that will stay dissolved in the soda depends only on how many collisions take place between carbon dioxide molecules and the surface of the liquid. If you had pumped in carbon dioxide gas, that would be another story; but nitrogen and oxygen are simply irrelevant.

Bottom line: keep it capped and keep it cold. It is especially important to keep the bottle tightly sealed while it is out of the refrigerator, because that's mainly when the carbon dioxide is emerging, due to the higher temperature. So pour what you want, cap the bottle, and put it right back in the fridge.

But don't get your hopes up too high. You can slow down the exodus of carbon dioxide, but you can't stop it.

And oh, yes. Whatever you do, never shake the bottle. That only speeds up the emergence of the gas.

TRY IT

Let a glass of cold water stand around for a few hours and as it warms up, you'll see bubbles of air forming on the walls of the glass. The air was dissolved in the cold water, but the warmer water can't retain that much air. It "goes flat," just like beer.

An Iconoclastic Elastic

Everybody knows that things expand when they're heated. But somebody wanted to bet me that there is a common substance around the house that will contract when heated. Should I have taken the bet?

The common substance is rubber. But only when it's stretched.

Most things expand when heated for a simple reason: the higher temperature makes the atoms or molecules move faster. They then need more elbow room, spreading farther apart on the average, and that makes the whole substance take up more space.

But rubber can behave differently because of its oddly-shaped molecules. They're like a can of worms—thin, squiggly-shaped chains, tangled all together into a disorderly snarl. That is, until you stretch the rubber. When you stretch it, the chains are stretched out, forced to line up along the direction of the stretch. But that is a very strained, unnatural state for them; you know that because you had to work to stretch them out that way, just as you would to stretch out a spring. As soon as you let go, the rubber molecules return to their compact, crinkled forms, and the rubber as a whole will snap back to its original shape.

What has that got to do with the effects of heat? Well, if you heat the rubber while it is in its stretched-out-molecule form, the agitation of the molecules makes them pull in on their ends, which tends

Bar Bet. Rubber can contract when heated.

That's stretched rubber, remember. Rubber that isn't stretched will expand when heated, just like anything else.

to decrease their length. (A wriggling snake is a shorter snake.) The rubber thus tries to revert as much as it can to its more compact form; it contracts.

TRY IT

Cut a wide rubber band, at least a quarter-inch wide, to make a strip, rather than a loop. Use a tan, rather than a colored, rubber band; the colored ones generally are not natural rubber. Tie a weight to one end of the strip and tack the other end to the edge of a shelf, letting the weight hang down freely. The weight should be heavy enough to stretch the rubber moderately. Now heat the rubber band with a hair dryer. Watch carefully and you'll see the rubber contract, pulling the weight up a bit higher.

Beating the Heat

How can the very same thermos bottle keep hot things hot and cold things cold, seemingly at our whim? Someone told me it's done with mirrors.

Part of the job really is done with mirrors. But it's not the thermos bottle that's playing tricks; it's your thinking.

To solve the problem, all you have to do is think of heat as a kind of liquid that flows only "downhill" from high temperatures to low temperatures. The thermos bottle acts like a dam that blocks the flow of heat. It won't let heat flow "down" from your hot coffee inside to the lower-temperature air outside. Similarly, it won't let heat flow "down" from the outside air to your lower-temperature iced tea inside.

Another way of saying this is that the walls of the thermos bottle are a very effective heat insulator, a substance or arrangement of substances that retards the flow of heat. We are most familiar with using insulators to keep heat from flowing out of our warm bodies and houses into the cold outdoors; ski jackets, sleeping bags, and attic insulation come readily to mind. But our refrigerators are also insulated, in this case to keep heat from flowing *in*. Insulators work both ways.

Heat, of course, is not a liquid, even though it does flow from one place to another. It moves in three ways: by conduction, by convection, and by radiation. Let's take them one by one and see how a thermos container foils them all.

Put a cool object in close contact with a warm one and you know what will happen: the warm object surrenders some of its heat to the cool one, so that the cool one becomes warmer and the warm one becomes cooler. Some heat has been transferred or conducted from the warmer object to the colder one. But what is heat, anyway? It is the agitation, or movement, of an object's molecules. The more vigorously its molecules are moving, the warmer it is. So when you place a warm object (having rapidly moving molecules) in close contact with a cooler object (having slowly moving molecules), some of the faster molecules will collide with the slower molecules, transferring some of their energy to them and speeding (warming) them up. That's *conduction*: direct molecule-to-molecule energy transfer.

When you touch a hot frying-pan handle, your skin molecules are speeded up by collisions with the frying pan's faster-moving molecules. When you touch an ice cube, your skin molecules lose some of their speed through collisions with the ice molecules.

A thermos container hinders conduction because it has double walls with nothing—a vacuum—in between. Because there are no molecules in a vacuum to collide with, heat conduction cannot take place through it.

Convection is the process whereby heat is transferred from one place to another by the actual bulk movement of a gas or liquid. You've heard people say that heat rises? What they really mean is that *hot air* rises, and along with it goes the heat it contains. That's *convection*. A convection oven is simply an oven with a fan in it that assists the circulation of hot air. In that case, the process is called *forced convection*.

A thermos bottle hinders convection simply by being a closed container; warm air or warm liquid can't pass through its walls. Any kind of closed container would stop convection.

Finally, heat can be radiated from one place to another in the form of infrared *radiation*. These energy waves are emitted by warm objects, fly through space, and can be absorbed by cooler objects, transferring their energy to them and heating them up.

A thermos container hinders infrared radiation by deflecting it with a mirror. The double walls of the container are silvered on their inner (vacuum-containing) surfaces, so any infrared radiation that tries to get through from either direction is reflected right back to where it came from.

If you think radiation is not a serious contender for heat transmission, consider how you broil a steak *underneath* the heating element in an electric oven. Heat goes upward by convection, all right, but a lot of it also goes downward (and in all other directions) by radiation.

No thermos container is perfect, of course. Some heat is always being conducted or radiated out of your hot coffee or into your iced tea. But the thermos slows down the heat-transfer processes substantially, and your food or beverage stays hot or cold for hours, rather than minutes.

Incidentally, the name Thermos (it's just the Greek word for "hot") started out as a trademark in 1904, but it became so widely used that

---

You didn't ask, but . . .
How does Styrofoam work as an insulator?

Unlike thermos, which has become a generic word, Styrofoam is still struggling to retain its name as a trademark, but nobody seems to be paying attention; people call all polystyrene foam products "Styrofoam" anyway.

The material is a good insulator because the plastic foam contains billions of trapped gas bubbles. Gases hinder heat conduction because their molecules are so far apart that they are very difficult for other molecules to collide with, either to give or to take away energy. The polystyrene plastic in between the bubbles is a good insulator also, because its molecules are so big that they cannot move around much.

The thin Styro—I mean, polystyrene foam—boxes that restaurants pack your doggie-bag food in are supposed to keep the food hot on your way home, as if that's when you were going to eat it. Instead of staying really hot, though, the food is probably being kept at just the right temperature for bacteria to flourish. Then when you get home, you put the whole box in the refrigerator for the next day's lunch, but the foam insulation may keep it at maximum spoilage temperature for another hour or so. Better to transfer the food into an uninsulated container before putting it in the refrigerator.

it's now a generic term for any vacuum container. One manufacturer still uses it as a brand name, however.

Freeze! I've Got You Uncovered

I took a can of soda pop from the refrigerator and the instant I opened it, it froze solid. What happened?

The soda wasn't frozen as long as it was still in the refrigerator because the refrigerator's temperature was warmer than its freezing point. But when you pulled the tab, you did two things: you released the pressure inside the can and you actually lost some of the gas. For different reasons, each of these effects helped the liquid to freeze.

Every liquid has a certain temperature at which it will freeze: its freezing point. The freezing point of pure water is 32°F or 0°C. Impure water—water that has any kind of stuff dissolved in it—has a colder freezing point than pure water does. The more stuff is dissolved in the water, the lower its freezing point will be.

Soda pop certainly has a lot of stuff dissolved in it: sugars, flavors, and especially carbon dioxide gas. So it won't freeze until well below 32°F. But as soon as you opened the can, the liquid lost some of its burden of dissolved carbon dioxide gas, which escaped from the liquid and went off into the air. Now containing less dissolved stuff, the liquid's freezing point became warmer than its own temperature from the refrigerator, and it dutifully froze.

Opening the can and releasing the pressure had another effect. Ice occupies more volume than liquid water does. So if you compress ice it tends to revert to its smaller-volume liquid state; it melts. Under the high-pressure conditions in the closed can, the ice was repressed and remained liquid. But as soon as you released the pressure, the liquid water was free to expand into its higher-volume form: ice. Of course, this couldn't have happened unless the soda was already colder than its freezing point because it had already lost the gas.

As if that were not enough, there is a third effect. When you opened the can, the compressed carbon dioxide gas was able to expand. Whenever a gas expands, it cools. This extra cooling also contributed to the freezing.

Either turn your refrigerator down—that is, turn the temperature up—or don't open the cans until they have warmed up a bit. You can wait.

Getting Hot in Bed

Why do water beds need heaters? A few days after filling one, won't the water be just as warm as anything else in the room, including any other type of bed?

Although several forms of water-filled beds were used in the early eighteenth century, the first US patent for "A Liquid Support for Human Bodies" was filed by Charles P. Hall in 1968 and waterbeds quickly became the passion of the hippie counterculture that prevailed at that time. By 1987, water beds comprised 22 percent of all mattress sales in the US. Sales have declined precipitously, but they are still widely available and a legitimate excuse for me to write about heat transmission.

The water in a waterbed—essentially a vinyl bag filled with water—will indeed settle down to the same temperature as everything else in the room, including a conventional mattress. But you would still feel colder on the waterbed. It has to do with the fact that water conducts heat away from your body a lot more efficiently than other materials do, such as the padding in a mattress.

Heat is nothing more than the motion of a substance's molecules. Various materials can transmit that motion, and hence conduct heat, with varying efficiencies. The best way is transmission directly from one molecule to the next to the next, and so on down the line. In order to do this, adjacent molecules must be close enough together to poke elbows. In water, the molecules are just about touching, so the faster-moving (hotter) molecules can easily transmit some of their motion to the adjacent cooler ones. The heat, in this case, your body heat, thus travels efficiently into the water, and you will feel cold unless some of that stolen heat is restored to you. Enter the electric heater.

Mattress padding is a much poorer conductor of heat than water is because it is essentially a bag of air, rather than water. In fact, a mattress is a very efficient heat insulator. The molecules in air are very far apart, with lots of empty space between them. They can therefore only rarely bump against each other, so heat transmission by molecular collisions is very slow. On a regular mattress, your body is putting out heat faster than the mattress can lead it away, so you stay cozy.

Want to be really cold? Try sleeping on a metal plate. Metals are superb conductors of heat because their atoms are cemented very closely together by a "cement" of electrons. That's why metals are great conductors of both heat and electricity.

But that's another story.

TRY IT

Try thawing two boxes of frozen strawberries, one by leaving it out in the air at room temperature, and the other by immersing it in a bowl of cold tap water. Even though the air is warmer than the water, the strawberries will thaw faster in the water because water conducts heat to the box—that is, removes cold from the box—more efficiently.

Bar Bet. You can thaw frozen strawberries faster at 65°F than at 75°F.

The Cigarette-Smoke Blues

I have heard that back in the Dark Ages when people actually smoked cigarettes, the smoke rising from the cigarette was blue. But after the doomed one inhaled the smoke and blew it out again, it was white. I know what probably happened to the lungs, but what happened to the smoke?

Tar and nicotine are not blue, so forget that idea. What happened was that the size of the smoke particles had changed.

The particles in cigarette smoke as it rises from a quietly burning cigarette are extremely tiny, smaller than the wavelengths of visible light. When a passing light wave encounters one of these tiny particles, the particle is too small to bounce the wave backward like a handball from a wall. Instead, the wave is merely deflected somewhat from its path and continues off at an angle: it is scattered by what scientists call Rayleigh scattering.

The shorter wavelengths of light at the blue end of the visible light spectrum are scattered more out of their original paths than the longer wavelengths are, because they are closer in size to the smoke particles. When we look at the smoke with the main source of light behind us or off to one side, many of the blue rays are not just going straight through and being "lost" to us; they are being scattered around the

You didn't ask, but . . .
No science book can be complete without answering the question,
"Why is the sky blue?"

Ho, hum. Well, it is blue for a similar reason that cigarette smoke is
blue: the preferential scattering of blue light by tiny particles. In a
non-polluted sky, the major tiny particles are the nitrogen and oxygen
molecules that make up the air.

Pure air is colorless, of course, meaning that all visible wavelengths
(colors) of light pass through it without being absorbed. But it contains nitrogen and oxygen molecules and often, suspended dust motes
that are much smaller than the wavelengths of visible light. As is the
case with the cigarette smoke particles, the blue, shorter-wave light is
scattered more than the other colors, which tend to go straight through
the air without much change of direction.

When you look at the sky, you are seeing all the colors in sunlight
that are coming straight toward you from the sun, usually from some
direction off to one side, wherever the sun happens to be. But in addition to that direct light, you're getting some extra blue light that comes
to you from many other directions after being scattered. Thus, you're
receiving an excess of blue light over what the sun is aiming straight
at you, and the sky looks bluer than the sun's own daylight.

You didn't ask this either, but . . .
Why are sunrises and sunsets so colorful?

At sunrise or sunset, the sun is low in the sky. To get to your eyes,
its rays must traverse a greater distance of atmosphere than when it
was more-or-less overhead. During its travel through all those miles
of atmosphere, a lot of the blue (short wave) light that started out in
your direction gets scattered by tiny dust particles into many other
directions, so the light that reaches you straight on is depleted in blue.
Sunlight that is depleted in blue looks red, orange, or yellow (longer-
wave colors), depending on what size the particles of dust in the air
happen to be.

If that kills the romance, forget that I ever said anything.

room more than the other colors. Thus, our eyes receive an excess of blue light and the smoke appears bluish.

When a cigarette is puffed upon, the smoke particles are somewhat bigger, because they don't get a chance to burn down completely. When inhaled, many of them become trapped in the lung, where they are not seen again until the autopsy. Those that do complete the round trip to lungland come out coated with moisture, which further increases their size. The particles are now bigger than the wavelengths of all colors of light, and they therefore don't scatter any of it. Like any large object, they reflect all colors equally, right back to where they came from. The smoke therefore does not appear to have any particular color and it looks white.

TRY IT

Make your own sunset. Add a few drops of milk to a clear glass of water and look through the glass at a light bulb. The bulb will look red, yellow or orange. The light coming to you from the bulb is depleted in blue because of scattering from the tiny casein particles and butterfat globules suspended in the milk. The exact color that you see depends on the size and concentration of these particles in the water.

The Fizz-ics of Champagne

Why does shaking a bottle of soda or beer make it explode when you open it? And does opening a bottle of champagne have to be so messy? After all, when it hoses out the candles, some of the romance is inevitably lost.

The trick, as you have already suspected, is to chill the bottles well and avoid any agitation for at least several hours before opening. But knowing why always helps.

Beer, soda, and champagne get their fizziness from carbon dioxide gas, which has been dissolved in the liquid during—or in the case of champagne, after—the bottling process. The fizz consists of bubbles of carbon dioxide coming out of the liquid into the air. When that happens gently on our tongues, we get that nice tingly sensation. But when it happens too fast, we get the mop.

The amount of carbon dioxide that can remain peacefully dissolved in a liquid depends directly on how much carbon dioxide there is in the space above the liquid's surface, because the more carbon dioxide molecules there are bouncing around in that space, the more of them will hit the surface and dissolve. In the sealed bottle, that space is filled with carbon dioxide and air; furthermore, these gases are packed in very tightly, at a pressure that can be as high as 60 pounds per square inch or 4.2 kilograms per square centimeter. (The air pressure in your tires is only about half that much.) So there is lots of carbon dioxide dissolved in the liquid when it comes from the bottler.

When the bottle is opened, no matter how gently, the pressurized carbon dioxide escapes and only normal air at normal pressure exists above the surface. In normal air, only about one out of every 3,000 molecules is carbon dioxide. So practically all of the dissolved carbon dioxide has to come out of the liquid in one way or another. The only question is, how fast?

After an initial burst of escapees, the rest of the dissolved carbon dioxide molecules cannot just leave the liquid all at once. If they did, your beverage would go flat as a shadow in an instant, making one hell of an explosion no matter how gently you had handled the bottle. Neither can the gas molecules leave one at a time from deep within the liquid; they have to find some rallying points, some unique meeting places at which they can congregate and form groups—bubbles—that are big enough to muscle their way up and out of the liquid. Scientists refer to these congregation sites as *nuclei* (plural of nucleus).

Almost any break in the homogeneity of the liquid, even a microscopic speck of dust, can serve as a nucleus for the formation of bubbles. So will tiny scratches on the surface of the glass, because they can trap microscopic air bubbles when the beverage is poured, and these air bubbles will invite more gas molecules to join them. Carbon dioxide molecules congregate at all of these nuclei and grow into bubbles, which rise as soon as they are big and buoyant enough to push their way upward through the liquid.

What does all this have to do with shaking the bottle? Well, when you shake the bottle, you are trapping some of the gas that was above the liquid, making tiny bubbles out of it. And these tiny bubbles are the best possible nuclei for the further growth of bubbles. The carbon dioxide molecules in the liquid latch onto these new bubbles and grow into bigger and bigger bubbles. Before you know it, you have a foaming

You didn't ask, but . . .
Why do dainty little bubbles form in a glass of champagne and rise in genteel little streams, while the bubbles in a glass of beer seem to belch up grossly from all over the place?

There are several reasons, none of them sociological.

(1) The champagne is likely to have been poured into a flute, a tall, narrow glass that does not have a lot of bottom surface for bubble formation. Moreover, such narrow glasses are not as likely to have been scratched on their inside surfaces because (a) scrubbing instruments can't get into them as easily and (b) they have probably been used less often than beer mugs. Fewer scratches mean fewer nuclei, which means fewer and smaller bubbles. You will see them rising from only a select few nucleation sites.

TRY IT   *. . . But scratch the inside of the glass with the tip of a knife, and you will see bubbles arising from that brand-new nucleation site.*

(2) Champagne is clearer than beer. True champagne (it says *méthode champenoise* on the label), as opposed to cheap sparkling wine, has been carefully clarified by cooling, settling, and *dégorgement* or disgorging. In that process, the corked bottles are tilted neck downward and rotated periodically over a long period; the neck is then frozen and the plug of frozen sediment is shot out with the cork. Less suspended matter in the liquid means, again, fewer nuclei for bubbles to grow on.

(3) The carbon dioxide in true champagne is made right there in the corked bottle by added yeast and sugar during an aging process that goes on for months and sometimes years. During that long time, the yeast cells not only die, as they do in beer and other wines, but their proteins decompose into fragments called peptides. Every peptide molecule has one end that is a base, which can grab onto a carbon dioxide molecule, which is an acid, thereby trapping it in the solution. So champagne cannot only hold more carbon dioxide than the other beverages, but it gives it up more reluctantly after the bottle has been opened. Hence, the tiny streams of aristocratic bubbles, rising in orderly, one-by-one fashion from whatever nuclei happen to be available.

If you stopper and refrigerate the bottle, good champagne will still be fizzy the morning after. And even the morning after that, if you are really serious about celebrating.

mess, propelled out of the neck of the bottle by expanding gas pressure like a pellet out of an air rifle.

You will run into the same problem, but probably not as bad, if you open beer, soda, or champagne that is not chilled enough. Carbon dioxide is less soluble in warmer liquids, so more gas will rush out than if the liquid were cold. If you have also shaken the bottle real good—well, DO NOT TRY THIS AT HOME!

This Spoon's Immune

At a dinner party at a friend's house, I stirred my coffee and the spoon got very hot, seemingly even hotter than the coffee. That never happens at home. What's going on?

Congratulations. Your friends think highly enough of you to put out their company tableware, which is made of sterling silver. Your home "silverware" is either stainless steel or (sorry about that) only silver-plated base metal.

Sterling silver is almost pure silver: 92.5 percent, to be exact. And silver is the best conductor of heat among all the metals. Heat will always move from a place of higher temperature to a place of lower temperature if it can find any way to get there, and silver provides a superb heat highway. All the spoon did was to conduct the coffee's heat out of the cup and into the cooler room, or when you touched it, into your fingers. During the process of being a conduit for all that heat, the spoon itself becomes hot: approximately the same temperature as the coffee, even though you might think it's hotter. (I don't recommend sticking your finger into the coffee to prove that.)

Stainless steel conducts heat less than one-fifth as well as silver does. At home, you probably never leave your everyday spoon in the coffee long enough for it to get very hot at the handle end. Even if you did, it wouldn't conduct its acquired heat into your fingers fast enough to be uncomfortable.

We All Scream for Ice Cream

My ice cream freezer uses an ice and salt slush to produce an extra-low temperature. How does salt make it so much colder than the usual temperature of ice water?

The normal temperature of an ice-and-water slush is 32°F (0°C). But that's not cold enough to freeze ice cream; it has to be at 27°F (−3°C) or below. Salt is what does the job. Many other chemicals would do the trick, but salt is cheap.

When ice and salt are mixed, some salt water is formed and the ice spontaneously dissolves in the salt water, making more salt water. That's what happens when you throw salt on an icy sidewalk or driveway; solid ice plus solid salt become liquid salt water.

Inside a piece of ice, the water molecules are fixed in a definite, rigid geometric arrangement. This rigid arrangement breaks down under

---

TRY IT   Put the same amount of cracked ice in each of two identical glasses. Pour just enough water into each one to make the ice begin to float. Then dump a lot of salt into one of the glasses and poke it down into the ice a bit. After several minutes, check the temperatures with a kitchen meat thermometer. (Lord knows why, but many of them do go down below freezing.) You will find that the salted ice gets much colder than the plain ice. You may even be able to scrape some frost off the outside of the salted glass with your fingernail.

the attack of the salt, and the water molecules are then free to move around loosely in the form of a liquid.

But it takes energy to tear down that solid structure of ice molecules, just as it takes energy to tear down a building. For a piece of ice that is in contact with nothing but salt and water, that energy can come from nowhere but the heat content of the salt water. So as the ice breaks down and dissolves, it borrows heat from the water and the temperature goes down. The slush is repaid by taking heat out of the ice-cream mixture, which is, of course, just what you want it to do.

Some Like it Hotter

Whenever I'm washing my hands, or worse yet when I'm taking a shower, I carefully mix the hot and cold water to get just the right temperature. Invariably, just as I'm getting comfortable, the water gets colder and I have to mix it all over again. Is there a scientific, rather than a paranoid-schizophrenic, answer to this?

Yes, and a very simple one. Heat makes things expand. In a compression faucet (the most common kind), the water flows through a narrow gap between a neoprene rubber washer and a metal "seat." In the hot-water faucet, the initial flow of hot water makes the washer expand, which closes down the gap between washer and seat, restricting the flow of water. With less hot water flowing than you originally selected, the mixture is now colder.

There are several things you can do:

(1) Replace the neoprene washer in the hot faucet with a "sandwich" type: fiber composite on the outside and rubber on the inside. The fiber doesn't expand and contract as much as rubber does.

(2) Don't be so stingy with the hot water. If you open the faucet wider, the slight constriction due to expansion won't even be noticeable. Of course, to get the temperature you want, you will have to open the cold-water faucet wider also.

(3) Preheat the hot water faucet parts by running the water for several seconds after it flows hot. Then when you adjust the temperature, the diabolical expansion will already have taken place.

(4) Take cold showers.

Or for a refreshing change of pace, ask your live-in to flush the toilet while you're in the shower. Instead of freezing, you'll be scalded.

Why Do Batteries Go Dead?

Almost everything these days is powered by batteries. What's inside them? It must be electricity in some form, but how does it stay in there until we want our bunnies to start going . . . and going . . . and going . . . ?

Batteries do not contain electricity as such; they contain the potential for electricity in the form of chemicals. These chemicals are isolated from one another inside the battery and are prevented from reacting until we hook the battery up to a device and turn on the switch. Then they react and produce electricity.

Getting energy from chemicals is nothing new. We get heat energy from wood, coal, and oil, chemicals all, by burning them—that is, by allowing them to react with the oxygen in the air. Batteries belong to a whole class of chemical reactions that can be made to give off electrical energy, rather than heat energy. They are *oxidation-reduction reactions*; chemists call them *redox* (REE-dox) *reactions* for short. They're quite common. Every time you use laundry bleach, for example, there is a redox reaction going on in your washing machine. You don't see the redox electricity because it is internal to the chemicals that are reacting; it is absorbed by certain molecules as fast as it is being produced by others. A battery is just a clever device that controls the chemical reactions in such a way that we can tap that electrical energy whenever we need it. But first, let's see exactly what electricity is.

An electric current is a flow of electrons from one place to another. But where do the electrons come from? Electrons are everywhere; they are the outer parts of all atoms. So if we want an electron to move from one place to another, it has to leave atom A and skip to atom B, like a flea jumping from dog to dog. In order for this to happen, however, atom A must be willing to give up one of its electrons and atom B must be willing to take it on.

Different kinds of atoms have different affinities, or strengths of attachment, to their electrons. Some atoms actually try to get rid of an electron or two whenever they can, while others hold onto their electrons tightly and will even try to capture more. When an atom of the first kind (atom A) meets an atom of the second kind (atom B), they can make a mutually beneficial deal by swapping an electron or two from A to B. And that, in a nutshell, is what happens in a *redox reaction*.

This electron-passing game from one atom to another constitutes a flow of electricity on a microscopic, one-atom-at-a-time scale. The problem from our human-sized point of view is that if we try to get a usable amount of electricity by mixing a zillion atoms of A with a zillion atoms of B, the electron-passing takes place from atom to atom in all directions, in one big chaotic scramble, wherever an A can find a B. That is of no practical use to us. We would like those electrons to be passed from a large group of A atoms in one location to a separate group of B atoms in another location, through a one-way street, or circuit, that we provide. Then, in their eagerness to get from the As to the Bs, those electrons will have to push their way through our circuit, doing work for us along the way, anything from lighting a flashlight bulb to making a little pink bunny wander vacuously around while beating on a drum.

To make a battery, then, we'll make a compact little package containing lots of A atoms and B atoms. But we'll keep them separated from one another, usually by a barrier of wet paper. They won't be able to do their electron passing until such time as we complete the circuit; that is, until we hook the battery up and close a switch that allows the electrons to flow from the A atoms through our interposed gadgetry to the B atoms.

Different types of batteries are made from different kinds of A and B atoms. The most common ones are manganese, zinc, lead, lithium, mercury, nickel, and cadmium. In the familiar AAA (no relation to what we've called "A atoms"), AA, C, and D batteries (there once was a B battery, but it isn't used anymore), zinc and manganese atoms are the As and Bs: the zinc atoms are the electron passers, and the manganese atoms are the electron receivers. The battery's *voltage*, in this case 1.5 volts, is a measure of the force with which zinc atoms pass their electrons to manganese atoms. Different combinations of passer and receiver atoms will make batteries with different voltages, because they have different amounts of eagerness for passing and receiving electrons.

When all the passer atoms have passed their quota of electrons to the receivers, the battery is dead, and alas! the bunny stops here. NiCad (nickel-cadmium) batteries, as well as your automobile's lead-acid battery, are rechargeable, however: we can reverse the electron-passing process by pumping electrons back from the receivers to the passers, and then the passing game can begin all over again. Unfortunately, though, every time the battery is recharged, some mechanical damage

You didn't ask, but . . .

Once a battery sends the electrons out into an electrical device, the electrons flow through the device and back to the battery again, don't they?

Not exactly. Inside the battery, electrons are indeed passed from one atom to another like jumping fleas. But that's not how electricity flows through a wire or through a complicated electric circuit. The electrons don't just enter one end of a wire, hop from one atom to the next, and come out the other end.

Let's say that the battery's voltage is pushing electrons through a wire from left to right. What really happens is that each electron repels its right-hand neighbor, because they are both negatively charged and same charges repel each other. This nudges the neighbor towards the next right-hand neighbor, which nudges its neighbor, and so on. By the time the wave of nudging gets to the other end of the wire— which is a lot faster than an electron can get there by broken-field running through the jungle of atoms—the effect is exactly the same as if those end-of-the-wire electrons were the original beginning-of-the-wire electrons.

Who can tell one electron from another, anyway? Not even another electron.

is done to its innards, and even a rechargeable battery will not last forever.

Out, Damned Spot!

How does laundry bleach tell white from colors? Apparently, it can take any stain that humans don't like, no matter what its chemical composition might be, and turn it into white. How does bleach know what we want it to do?

Bleach knows nothing about white. What it does know about is color, because color is a lot more fundamental, chemically and physically speaking, than our mere human washday preferences. Bleach attacks colored chemical compounds, most of which do indeed have something in common, and leaves behind a lack of color that we like to think of as "white."

Before I myself am attacked for calling white the absence of color, when you learned in school that white is the presence of all colors, let me explain.

Light from the sun does indeed contain all colors of the rainbow: all the colors that humans can see, and then some. When all of these colors of light are combined, as they are in daylight, our particular brand of vision perceives the light as being no particular color at all. We call it white light.

But that's the light itself. What do we see when we look at an object being illuminated by that light? If the object reflects back to our eyes, equally, all the colors that fell upon it in the form of daylight, then the reflected light still has no apparent color to us; it is still white. We say that the object itself is white, because we can only judge it by the light that it sends to our eyes. Beauty is in the light that reaches the beholder.

If, however, the object has a particular appetite for, let's say, blue light, and it absorbs or holds back some blue parts of the daylight before reflecting the rest back to us, then the light we see will be deficient in blue, which our eyes may perceive as yellowish, and we therefore call the object itself yellowish.

If the "object" in question happens to be a stain on our otherwise-white (colorless) shirt, we turn it over to our good old dependable laundry bleach for obliteration. We'll do the same when a stain happens to absorb some other specific color of light, thereby appearing to us as some other non-white color.

What is it, then, that the bleach is actually acting upon when it removes color? It is acting upon those molecules that prefer absorbing some specific color or colors of light. Any specific color. The question then becomes, how does bleach attack only light-absorbing molecules?

When a substance absorbs light energy, it is the electrons in the molecules that do the absorbing. By absorbing the energy, the electrons promote themselves to a higher level of energy-status within their molecules. The molecules of many substances are colored because they contain electrons that are of particularly low energy-status to begin with, and that are therefore eager absorbers of light energy. What bleach molecules do is gobble up these low-energy electrons so that they are no longer available to absorb light; thus, the molecules lose their coloring ability. (Electron gobblers are called oxidizing agents; the bleach oxidizes the colored substance.)

The electron gobbler that is commonly used in the laundry is sodium hypochlorite. Liquid bleaches are nothing but a 5.25-percent solution of that chemical in water. Powdered bleaches are usually sodium perborate, a gentler electron gobbler that does not attack most dyes. (Dyes, of course, are nothing but deliberate, tenacious, light-absorbing stains.)

Another popular electron gobbler, hydrogen peroxide, is used to bleach melanin, the dark coloring matter in human hair and skin. It is widely used in the manufacture of blondes.

Whoops!

Why is it that when I drop something on the floor, a wine cork, say, or a screw or some other small object, I can never find it where I expect it to be? Even if I saw it hit the floor, when I look for it later, it is never anywhere near that spot.

Newton's Fourth Law of Motion says that a body released in free fall will come to rest in as inaccessible a location as possible, preferably beneath a heavy piece of furniture or down a nearby drain.

Okay, so Newton didn't say that. Maybe it was Galileo who, as legend has it, enjoyed dropping things off the Leaning Tower of Pisa. But historians are not certain that he actually did that experiment, rather than figuring it out in his head, so let's drop the whole thing.

With the exception of meatballs, which despite their name do not roll and are disasters on golf courses and bowling alleys, a dropped object will not stick to its point of impact with the floor. (We're assuming a hard floor, like wood or vinyl. Not carpet.) It will hit with some unpredictable amount of angular momentum, depending on what physicists could call its *moment of clumsiness*: the fumbling that takes place just before the drop. Fumbling is notorious for being erratic in its directional impulses; just ask any football pass receiver. So during its descent, the object is undoubtedly rotating and tumbling like Jack and Jill. In other words, it has *angular momentum*.

Collisions with hard floors are *elastic collisions*, meaning that the falling object retains all its kinetic energy, its energy of motion, unless it gives some of it to the object with which it collides. (Notice that the floor doesn't move.) But the object hits the floor while tumbling in some direction or other. As it bounces away, it retains its full complement of angular momentum, because obviously the floor doesn't begin to rotate. In scientific lingo, *angular momentum is conserved*.

You didn't ask, but. . .
What if the dropped object is food? Does the five-second rule still hold, even if it rolls around before coming to rest?

The "rule" says that if you pick it up within five seconds, it will still be safe to eat. The thinking, if I may so flatter it, is that pathogenic microorganisms on the floor need time to discover the food and set up housekeeping on it, so we can foil them by snatching the food away before the entire herd has had a chance to hop aboard.

That is simply absurd. Several hundred years of microscopic examination of bacteria have failed to detect any tiny trampolines or pogo sticks with which they can jump onto a piece of food. They don't hop, or even crawl. They live in the dirt, and dirt doesn't travel.

Thus, the length of time the food rests on the floor has nothing whatsoever to do with how contaminated it gets. Time is a totally irrelevant variable. The only relevant variable is stickiness; a pre-sucked hard candy, for example, will pick up more germs than a raw kumquat would. So you can pick up the kumquat when you find it under the table, brush it off on your shirt, and pop it into your mouth without concern.

The pre-sucked hard candy? It can be dangerous, no matter how quickly you pick it up.

Depending on what part of the object (assuming it is not a perfect sphere) impacts the floor, it will bounce off in a direction that is completely unpredictable, without knowing all the details of its shape, speed, and contact angle and feeding the data into a supercomputer.

Under such circumstances, do you really expect to be able to predict how far and in what direction it will skid, roll, or tumble? It could be miles—well, yards—from where it hit the floor. You just can't win this game, unless you play it with meatballs.

Oh, and another principle that Newton noted but was unable to explain was that dropped objects apparently develop some sort of GPS while falling, so they can seek out the darkest and most inaccessible cranny in the room to hide in.

Scientists are still working on that one. Stay in touch.

CHAPTER 4

The Infernal Combustion Engine

Your car is rusting away before your eyes, it will not start, your tires are flat, there is a bull's-eye rock chip in the windshield, and you are skidding down your icy driveway towards your concrete mailbox post. Wouldn't it be comforting to understand what is going on behind these events? Well, maybe after you've calmed down.

Here is where we look at some of the fascinating and frustrating phenomena that have grown out of our love affair with the infernal combustion engine.

Current Events

In cold weather, my car's battery acts half dead. In really cold weather, it won't even start the car. Yet I've been told to keep my flashlight batteries in the refrigerator to keep them lively. Why is cold good for flashlight batteries, but bad for car batteries?

Nobody is telling you to try to use the flashlight batteries when they're still cold. They'd be just as sluggish as your car battery; cold inhibits both kinds. They must be somewhere around room temperature if you are to get the intended amount of "juice" out of them.

Batteries produce electricity by a chemical redox reaction, and all chemical reactions go slower at lower temperatures. Cool a battery much below normal room temperature and the number of electrons it can put out per second (the amount of current it can deliver, expressed in amperes) is severely limited, whether it is in your car or in your flashlight.

It is only the battery's ability to deliver *current*—streams of electrons—that is inhibited by cold temperatures. Cold has virtually no effect on the force—the *voltage*—with which the battery sends out those electrons.

Another thing: batteries leak a bit of electricity even when they are not hooked up, that is, even when they are not delivering electricity. This leakage current eats into their limited supply of chemicals. If you keep them cold, you're slowing down even this small amount of chemical reaction and preserving the power for when you really need it. But today's alkaline batteries have such a long shelf life that refrigeration will hardly make a difference.

In automobile batteries, which contain a liquid (sulfuric acid), there is another cold-restriction factor. When the battery is delivering current, certain atoms (actually, they are positively charged ions) must migrate, or swim, through the acid from the positive internal pole to the negative pole. At cold temperatures, they are substantially slowed down, so the battery's ability to deliver current is also inhibited.

Some old-time garage mechanics will swear to you that if they leave a car battery for a long time on the concrete floor instead of on a shelf, the concrete "sucks the electricity out of it." What's going on, of course, is that the floor is cold and sucks the heat out of it.

What you really have to watch out for is the mechanic who sucks the money right out of your wallet.

Shatters Matter

For obvious reasons, automobile windshields are made so that when they break, they don't fall apart into a million pieces. But the other windows do break into tiny pieces. How do they get glass to break in these two different ways?

It is relatively easy to prevent the scatter of fragments from a broken windshield. It is constructed as a three-layer laminate: a sandwich of two glass "bread slices" with a tough, plastic "ham" in between. When a stone or a driver's head—forgive the tactless image—hits the windshield, it (the windshield, that is, and hopefully also the head) stays in one piece because the cracked glass remains bonded to the plastic layer. And usually, because the whole thing is glued to the car's frame, the entire cracked windshield will remain in place.

The side windows are not in as much risk of being broken by a hurtling object, so they don't have to be kept from shattering into flying bits of glass. But when they do, the bits must not be dangerously sharp, as when ordinary glass breaks. This is accomplished by *tempering* the glass—strengthening it—before it is installed.

To make a material stronger, engineers often resort to *prestressing* it: subjecting it to a carefully applied stress and then locking that stress energy into the final product.

Auto glass is prestressed by heat-tempering it. After the glass has been formed into shape, and while it is still very hot, the surfaces—and only the surfaces—are instantaneously chilled. This freezes in the molecular structure of high-temperature glass, which has a more expanded structure than room-temperature glass. When the whole sheet is then allowed to cool slowly down to room temperature, it retains the high-temperature structure frozen into its skin, while the interior shrinks down to the tighter room-temperature structure.

Thus, a combination of opposing tension (pulling) and compression (pressing) forces has been locked into the product as a pent-up push-me-pull-you energy that strengthens the entire structure.

The pent-up energy is released the instant the glass becomes flawed or cracked anywhere. Utilizing this energy, the fracture quickly spreads like a chain reaction over the stressed surface. Because every part of

---

You didn't ask, but . . .
How do they make prestressed concrete?

They play loud, hard rock music while it's being poured.
    Sorry.
    The strength of prestressed concrete doesn't come from heat tempering, as in the case of auto window glass.
    Concrete is very strong under compression, but under tension it is hardly worth its weight in taffy. Prestressing increases its tension strength.
    The concrete mix is poured over steel cables that are being pulled lengthwise, as if in an attempt to stretch them like rubber bands. Their stiffness fights against this pulling; they "want to" contract as if they were indeed rubber bands, but since their ends are tied they cannot release their tension that way. So they hold their tension and it is transferred to the concrete as it hardens. This stress energy is now locked into the concrete structure, pulling it together and making it more resistant to stretching and bending forces.

the surface is stressed, the cracks and breaks erupt equally all over it, resulting in a million gravel-sized pieces instead of a smaller number of dangerous shards.

The tempered surface of the side windows is so strong that you probably can't even smash it with a hammer. Auto accessory stores sell an emergency window-breaker hammer that has a conical, steel head. The point of the cone concentrates the hammer's force in a small spot on the glass, essentially pricking the hard, tempered skin of the glass, and allowing the window to fall apart.

Busting the Rust Trust

Everything I have is rusting away. Well, not really, but it seems I am always fighting off rust by oiling, scraping, and painting every- thing I own, from wrenches to lawn mowers and porch railings. I won't even mention automobiles. Maybe if I knew more about what causes rust in the first place, I could head it off. Any help?

Iron plus oxygen plus water equals rust. Or, in chemical lingo, $Fe + O_2 + H_2O \rightarrow Fe_2O_3$. That's it. When all three of these reactants are present, rust will inevitably occur. But if any member of this unholy trinity is missing, there can be no rust.

Fortunately for us living creatures, but unfortunately for our garden tools and automobiles, oxygen and water vapor are present everywhere in the atmosphere. And fortunately or unfortunately, the entire center of our planet, a core that measures some 4,000 miles in diameter, is almost 90 percent iron. Even the sun and the stars contain iron.

Here on the surface of the Earth, from whence we dig our miner- als, iron is the most abundant of the 88 known metallic elements. It is therefore the cheapest of all metals and the most widely used, whether in the form of wrought iron, steel (iron with carbon in it), or any one of dozens of alloys.

You just can't get away from iron, oxygen, and water, so it's no wonder you have a problem. But you're not alone. The rusting of iron has plagued humankind since prehistoric times.

The main villain is oxygen. In a process called oxidation, oxygen reacts with most metals to form oxides, and rust is a form of iron ox- ide. (In chemical circles, it goes by the name of hydrated ferric oxide.) Under the right conditions, oxygen will also react with aluminum, chromium, copper, lead, magnesium, mercury, nickel, platinum, silver, tin, uranium, and zinc, and many others.

You didn't ask, but . . .

Why does salt, whether it is present in the air near the ocean or whether it is being used on highways during freezing winters, make cars rust faster?

Rusting takes place through a juxtaposition of iron and oxygen that actually constitutes a miniature electric battery, on the atomic scale. That is, the oxygen molecules are taking electrons away from the iron atoms, and that is exactly what goes on inside a battery: electrons being snatched from one substance by another. Anything that helps electrons to get from the iron atoms to the oxygen molecules will help this process along.

Salt helps, because when salt dissolves in water it makes a solution that is a good conductor of electrons. Therefore, salt helps iron to rust by helping to deliver the iron atoms' electrons to the voracious oxygen molecules.

In fact, among all the metals you are probably familiar with, only gold is completely immune to its attack. That fact, plus gold's scarcity and unique color, is what makes it so highly prized.

(Incidentally, those jewelry-cleaning products that claim to "remove tarnish" from your gold jewelry are a fraud. Gold doesn't tarnish. Plain soap and water will remove any dirt.)

Oxidation doesn't corrode, deface, and destroy any other common metal the way it does iron. That's because most other metals have some sort of saving grace that keeps the oxygen from chewing it up. For example, oxygen reacts very readily with aluminum, but it happens that the first thin layer of oxide on the metal's surface is so tough and airtight that it seals off the rest of the metal from further attack. Other metals, such as copper, for example, react so slowly that all they do is darken a bit. The coating of oxide then protects the rest of the metal from severe corrosion.

When oxygen and water attack iron, however, the reddish-brown oxide doesn't stick. As you know from sad experience, it tends to flake off and crumble away, uncovering ever-fresher surfaces of metal for the air and moisture to ravage. Iron oxide's molecular arrangement just happens to make it a weak and crumbly material, and there is nothing we can do about it. There are products on the market, however, that

Nitpicker's Corner. In the rather complex atom-by-atom mechanism of rusting, salt helps also to conduct charged iron atoms (ions) to where they need to go. Moreover, the chloride in the salt, which is sodium chloride, has a separate effect on the iron. But that's all a bit more than we want to get into. Trust me. Just don't drive your car in salt water.

can convert the structure of rust into a tough, adherent coating. Check your hardware store.

The only lines of home defense against rusting, then, involve keeping the iron away from prolonged contact with moisture or oxygen. Never put your tools away wet. And anything that will fit into an airtight plastic bag can rust only as long as the limited amount of oxygen in the bag holds out. Sorry, but that's about all you can do, short of painting.

TRY IT

Even when immersed in water, iron won't rust if there is no oxygen present. Boil some water vigorously for several minutes to get most of the dissolved air out of it. Fill a jar with the boiled water and let it cool to room temperature. When it is cool, fill a similar jar with fresh tap water. Drop an iron nail into each jar and wait a couple of days. The nail in the boiled water will rust a lot less than the one in the tap water. (Boiling can't remove every bit of oxygen.)

Help! My Antifreeze Froze!

Expecting an unusually cold winter, I drained my car's cooling system and put straight antifreeze in it instead of the usual fifty-fifty mixture with water. Now a mechanic tells me that straight antifreeze freezes at a warmer temperature than 50-percent antifreeze does. How is that possible?

Strange as it may sound, your mechanic is correct. A fifty-fifty mixture of ethylene glycol and water will not freeze until the temperature gets down to $-34°F$ ($-37°C$), while pure antifreeze will freeze at the much higher temperature of +11°F (−12°C). Let's see what's going on here.

It happens that mixing almost anything at all into water will lower its freezing point. In principle, you could add salt, sugar, maple syrup, or battery acid to your engine coolant to lower its freezing point, and they would all do so, but for obvious reasons they're not recommended.

In the earliest days of automobiles, people did occasionally use sugar and honey as antifreeze. Later, alcohol became popular, but it boils off too soon. These days, we use a colorless liquid chemical called *ethylene glycol*, which is not volatile and doesn't boil off. Commercial antifreeze also contains rust inhibitors and a vivid dye to help us locate leaks in the cooling system and, not incidentally, to make it look technologically sophisticated.

The freeze-protecting power of dissolved substances in water has to do with a fundamental difference between the arrangements of the molecules in liquids (such as water) and in solids (such as ice).

In water, as in all liquids, the molecules are slithering freely around like a mass of oiled bodies at an orgy. They're loosely attracted to one another, but they're not connected in fixed positions, as they are in most solids. That's why you can pour a liquid, but not a solid.

You didn't ask, but . . .
The label on the antifreeze container says that it not only keeps the coolant from freezing, it also keeps it from boiling. How is boiling related to freezing?

By interfering with water's molecules, dissolved substances not only lower its freezing point, they also raise its boiling point by making it harder for the water molecules to fly off into the air. So with ethylene glycol dissolved in it, your coolant water has to get to a higher temperature than usual before it will boil. A mixture of 50-percent ethylene glycol in water won't boil until 226°F (108°C). That's less of an advantage than it used to be, however, because today's cooling systems are pressurized, and at elevated pressures the boiling points of both water and ethylene glycol are already higher than they would be at atmospheric pressure.

In order for liquid water to freeze, then, the molecules must slow down and settle into the highly proper, rigid positions that they must occupy in a crystal of ice. If given enough time to find these positions—that is, if the molecules are slowed down gradually enough by gradual cooling—water is capable of forming rather large chunks of ice. And that is exactly what we are afraid of, because when water freezes, it expands and the consequent pressure can crack the cooling passages in the engine block.

Alien molecules in the water, such as ethylene glycol, for example, throw a monkey wrench into this freezing process in two ways. First, by simply cluttering up the place, they interfere with the water molecules' ability to fall into those precise locations that are needed to form a crystal of solid ice. It's as if a military drill team were trying to fall into formation while a mob of civilians is running around on the field. By getting in the way, the alien molecules prevent the ice crystals from growing as large and uniformly as they would like to. Even if the water does freeze, then, the result will be a slush of small ice crystals, rather than a single, rock-hard, engine-cracking iceberg.

But the main effect that extraneous molecules have on freezing water is that they keep the water from freezing until a lower-than-normal temperature. What happens is that the ethylene glycol molecules are "diluting" the water, thereby reducing the number of water molecules that can congregate in any one spot to form an ice crystal. Because of that, we have to slow down even more of the water molecules, by lowering the temperature even more, in order to get enough of them to fall into place together as an ice crystal.

Why then will pure ethylene glycol freeze at a higher temperature than a 50-percent mixture with water? It is because ethylene glycol's molecules are interfered with by water molecules, just the same as water's molecules are interfered with by ethylene glycol molecules. It works both ways. The water lowers ethylene glycol's freezing point, even as the ethylene glycol lowers the water's freezing point. So ethylene glycol mixed with water will not freeze as easily as it did when it was pure.

Yes, it can be said that water keeps antifreeze from freezing.

Bar Bet. In your car's cooling system, straight antifreeze will freeze sooner than a mixture of antifreeze and water.

Car-Skiing: A One-Time-Only Sport

I live in a cold climate, and my house has a steeply sloped driveway. When the driveway is icy, I sprinkle sand on it to improve the traction of my tires. But the last time I tried it (and it will be the last time), the sand didn't work. It acted like so many tiny ball bearings under my tires, with highly unpleasant consequences. Why didn't the sand give my car traction?

It was an extremely cold day, wasn't it? Below zero Fahrenheit, perhaps? That was the problem. Sand will not work when it is too cold.

In order to improve traction, the sand grains must become partially embedded into the ice, making tiny bumps in what had been a smooth surface—in effect, making "sandpaper" out of the ice. It is the pressure of the car on the sand that accomplishes this. When the tire presses a sand grain against the ice, a bit of the ice melts beneath the grain, and it sinks in. The water then refreezes around the grain.

The ice melts under this pressure because ice is the bigger-volume form of water, and when pressed upon, it reverts to its smaller-volume form: liquid water. Without this pressure-melting effect, the sand could not embed itself into the ice.

The problem is that the colder the ice is, the more pressure is needed to melt it, because the water molecules in the ice crystal are more rigidly fixed in place and cannot easily be persuaded to move around loosely, as the molecules of a liquid do. Even though a car applies a lot of pressure to a grain of sand, it may not be enough to melt the ice in subzero weather.

You might do better on foot. A rubber tire is not the greatest pressure-applying device, because of its elasticity. The soles of your shoes are probably harder, and even though I presume that you weigh less than one-fourth as much as your car (one wheel's worth), you may still be applying more pressure to the sand grains—more pounds per square inch of grain—than the car does, and the sand will embed itself by the melting mechanism.

The Salt Man Cometh

When there is ice on my driveway, I throw salt on it and the ice melts. But how can anything melt ice without heat? They say it's because salt lowers the freezing point of water, but what can that possibly mean to the ice? It's already frozen.

Contrary to what everybody says, the ice on your driveway does not melt, any more than sugar melts in coffee or tea. People often confuse *melting* with *dissolving*. ("I don't need an umbrella; I won't melt in the rain.") But melting, as you have already noted, requires heat. You can certainly melt ice or sugar by heating them, but that's not what the salt does to the ice. The salt *dissolves* the ice. People use the word "melting" for the salt-on-ice phenomenon only because they see ice disappearing and a liquid—salt water—remaining. And "melt" just happens to be the word that our ancestors invented for "ice go 'way, water come."

Science teachers and textbooks, however, should know better than to fall into that linguistic trap. Along with many others, you were probably taught in school that "salt lowers the freezing point of water." But that's not literally true either. Throwing salt onto your driveway cannot possibly change the freezing point of water—the temperature at which good old H_2O is accustomed to freezing or melting. That temperature—it's the same for melting as for freezing—is 32°F or 0°C; it always has been, and it always will be. What the textbooks and teachers should be saying is that salt water freezes at a lower temperature than pure water does. That is quite a different statement.

On your driveway, the salt first makes salt water out of the ice, and then the resulting salt water remains unfrozen because its freezing point—not *pure water's* freezing point—is indeed below the temperature of the air. A fine distinction, perhaps, but critical to understanding what's going on.

First, how does the salt make salt water out of the ice? It happens that sodium and chlorine atoms (actually, sodium and chloride ions, but we won't quibble), which make up the sodium chloride or salt, have a strong affinity for water molecules. (Manufacturers have to add an anticaking agent to keep table salt from gumming up in the shaker because of moisture absorbed from the air.)

When a crystal of salt lands on an ice surface, the salt's sodium and chloride ions pull some water molecules out of the surface; they then proceed to dissolve in that water to form a tiny puddle of salt water

Bar Bet. Salt does *not* melt ice.

around the crystal. The puddle of salt water doesn't freeze because its freezing point is lower than the temperature of the air.

The sodium and chlorine atoms that are now dissolved in the salt water keep nipping away at the ice surface like piranhas going after a meatball in a punch bowl. As the process continues, more ice continues to dissolve in the salt water, making more and more salt water.

Eventually, either all the ice runs out, or the puddle of salt water gets so dilute that its freezing point is no longer below the air's temperature, and it will freeze. But salt water freezes only into slush, rather than into hard ice. In either case, your ice-destroying mission has been accomplished.

Now go tell your friendly neighborhood chemist that salt *doesn't* melt ice, and be prepared to duck.

The Duck's-Back Caper

Why won't oil and water mix?

Ordinarily, water is the best mixer in the world, and I don't mean just with Scotch. It mixes with, that is, it snuggles intimately with, it even welcomes into its very bosom—that is, it *dissolves*—more substances than any other liquid. It is sometimes called the *universal solvent*. But there is one family of substances that water abhors and will invariably shun: oils. Water won't even nuzzle up close enough to a drop of oil to wet it, much less to dissolve it. Water rolls off a duck's back because the duck's feathers are oily, and they don't even get wet when the duck goes diving. But you knew that.

Like guests at a social gathering, molecules must have at least something in common in order to mix successfully. Quite literally, the molecules of water and oil have practically nothing in common. Water, as you may know, consists of small, three-atom molecules: two hydrogen atoms and one oxygen atom. Oils, on the other hand, are made of big molecules consisting of many carbon and hydrogen atoms, with no oxygen needed at all. No matter how intimate the gathering, it is not very likely that the twain are going to meet and form an alliance.

What is there about oils that makes them outcasts in the big, wide, wonderful world of water, the most plentiful liquid on earth? Once we see why water is such a powerful solvent (dissolver) for so many other substances, we'll see that oils simply don't have what it takes to dissolve in water.

In pure water, as in any liquid, the molecules are being held together by some kind of mutual attraction. If they weren't, they would go flying off into the air and the liquid wouldn't be a liquid anymore; it would be a gas. The attractions between the molecules in water are rather special; they come from the fact that water molecules are *polar*: they are like tiny bar magnets, but instead of having north and south magnetic poles at their opposite ends, they have positive and negative *electric poles*—that is, positive and negative electric charges.

Now if you think of a glass of water as a glass full of tiny magnets all stuck together, you can see that they would have precious little interest in associating with any substance whose molecules weren't also magnets. Magnets are attracted to nothing but other magnets. (Yes,

Nitpicker's Corner. Besides the attraction of polar molecules, or "electric magnets," for one another, there's another important kind of attraction between water molecules. It's called *hydrogen bonding*. Without going into detail, let's just say that it can come about when molecules have an oxygen atom plus a hydrogen atom—a so-called hydroxyl group, OH—at one end. Water molecules fit this prescription precisely, and they stick together by hydrogen bonding as well as by polar attraction.

On the "like dissolves like" principle, other substances that are good setups for hydrogen bonding should also be likely to dissolve in water. And they are. Sugar (sucrose) is a familiar example. It dissolves in water, not because its molecules are polar, which they're not, but because they contain the water-like hydroxyl group and therefore can form hydrogen bonds to water molecules. The sucrose molecule actually contains eight hydroxyl groups.

If oil molecules aren't polar, and if they don't form hydrogen bonds, then what holds them to each other? It's a totally different kind of molecule-to-molecule attraction called a *van der Waals attraction*, about which we needn't bother our heads. Suffice it to say that these attractions are just as alien to water molecules as electric poles are to oil molecules. The revulsion of water for oil, then, is quite mutual.

a magnet is attracted to a piece of ordinary iron, but inside that iron are zillions of tiny magnetic regions called *domains*.

Only if a substance contains atoms or molecules with *electric* poles, then, will water be attracted to it, first by wetting it and eventually by enveloping it and dissolving it. Lots of substances fit this bill and will mix with water, but oils absolutely don't and won't, because there is nothing at all polar—no electric poles—in those big, long molecules of oil. So there is nothing that might be attractive to a water molecule.

Dissolving is the most intimate possible kind of mixing: the molecules of one substance mix, one-by-one, in amongst the individual molecules of the other. Where dissolving is concerned, then, we might conclude that only birds of a feather are likely to want to flock so close together. Being less poetic, chemists prefer to say that "like dissolves like," meaning that only substances having molecules similar to those of water are likely to mix with water. And ditto for oil-like molecules and oil. Generalizing even further, we can expect that a given substance, if it dissolves in anything at all, will dissolve in either oil or water, but not both. And that expectation is generally borne out. Salt and sugar dissolve in water; gasoline, greases, and waxes dissolve in oils. But never the other way around.

Many a Slip

Why is oil so good at lubricating things?

Obviously, because it's so slippery. But what makes a substance slippery?

All liquids are inherently slippery to some degree. A wet floor or highway is a well-recognized hazard that keeps lawyers in expensive clothing. But water isn't much use as a lubricant in our engines and other machinery because it is really not all that slippery and it evaporates away.

Oil is much slipperier than water because its molecules can slide past each other more easily than water's molecules can. And because a liquid is nothing more than a pile of molecules, when the molecules slide, you slide. You wouldn't be surprised to slip on a pile of ball bearings, would you?

Water molecules don't slide around as easily as oil molecules do, because they have a significant amount of stickiness—attractions to each other. Water's particular type of molecule-to-molecule attractions

Nitpicker's Corner. Oil molecules have to come up with some other way of sticking together, because if they didn't stick together at all, they would go flying off into the air as a vapor and all the machinery of civilization would grind to a screeching, smoking halt.

Oil molecules stick together by what chemists call *van der Waals attractions* or *van der Waals forces*. They explain these forces by waving their arms around a lot and muttering about electron clouds. The story goes that when a bunch of atoms cluster together to make a molecule, they pool their electrons to form a big, squishy cloud of electrons that swarms around the whole molecule like a horde of fruitflies around a cluster of grapes. So when two molecules come together, the first thing they see is one another's electron clouds. Flies meeting flies.

So far, so good. Nobody argues with that picture, which has served chemists extremely well in explaining how molecules interact with one another. But now for the weird part. In spite of the fact that all the electrons in these clouds have the same charge (negative) and should therefore repel one another, they supposedly somehow attract one another to hold the molecules together. That's what Professor van der Waals said, and he got a Nobel Prize for it in 1910. So go argue.

Anyway, these van der Waals forces do hold oil molecules together—especially the bigger oil molecules that have large electron clouds—strongly enough so they don't easily evaporate away from one another. That's why oil puddles don't "dry up" as water puddles do. But because they're joined together only by the squishiness and mushiness of electron clouds, the molecules can slide easily past one another.

(*hydrogen bonds*) arises predominantly in molecules that contain oxygen atoms, as water molecules certainly do. In fact, a water molecule is practically nothing but an oxygen atom. If we could see a water molecule, it would look like a big, spherical oxygen atom with two tiny hydrogen atoms glued on like a couple of peas on an orange.

But oil molecules—the molecules of *hydrocarbons* that make up that gooey, black mish-mash called petroleum—are made up of nothing but hydrogen and carbon atoms. No oxygen atoms at all. They therefore don't stick together very well and can slide easily over and under one another. Hence, they are good lubricants.

Pumping Irony

How come I can pump my bicycle tires up to 60 pounds in no time, but I have to knock myself out with the bicycle pump just to add a couple of pounds to my car's tire, which is only inflated to 30 pounds?

It's not just the pressure you're fighting against; you also have to consider the volume. It takes a lot more strokes of the pump to add a "pound of air" to the car tire than to the bike tire.

What people loosely call a "pound of air" is not an amount of air, like a pound of butter; it's a pressure: *pounds of force per square inch*, generally abbreviated *psi*. That force is the cumulative effect of the zillions of air molecules in the tire, which are continually bombarding every square inch of the inner walls. The more air molecules you force into a tire, the more bombardment there will be and the higher will be the pressure. Adding more air increases the pressure.

As you have surmised, it should be harder to force air into a 60 psi tire than into a 30 psi tire. That's because the air molecules inside a tire are also bombarding the valve opening, making it harder to force more molecules through. So each stroke of your pump does indeed require more effort to overcome the bike's 60 psi of pressure than to overcome the car's 30 psi. You have use twice as many pounds of force on the pump handle to push air into the bike tire.

Then why is it more work to pump up the car tire?

A typical car tire contains about four or five times as much air space as a typical bike tire. In order to have the same pressure—the same rate of molecular bombardment per square inch—in both tires, there would have to be four or five times as many air molecules present in the car tire. Therefore, to increase the car tire's pressure by each psi, you have to pump in four or five times as much air, using four or five times as many strokes, as you do to get a psi of pressure into the bike tire. Even though each stroke takes half the effort, you're still working more than twice as hard.

Inflation Is Heating Up

When I inflate my bike's tire with a bicycle pump, the tire gets hot. I assume that it's from the friction of all that air squeezing through the narrow valve. But when I fill the same tire at the gas station, the valve doesn't get hot. What gives?

It can't be friction, because approximately the same amount of air is being forced through the valve in both cases. The answer is that when air (or any gas) is compressed—that is, when it is forced into a smaller space—it gets hot.

When you use your hand pump, you're compressing the air in the pump's cylinder, but when you use the gas station's air, you're using air that has already been compressed. The gas station's air did indeed get hot when it was originally compressed into the storage tank. But by the time you show up with your sad-looking tires, the air has had lots of time to cool off. All you were doing was bleeding off some of that stored-up air. No compression was going on, so there was no heat.

Why does compressing a gas make it hot? Well, gas molecules are free spirits; they are flying around freely, as far apart from one another

---

You didn't ask, but . . .
If compressing air makes it hotter, does air get cooler when it expands?

Definitely. And that is indeed what is happening back in the gas station's compressed-air tank as you allow some of its stored-up compressed air to expand into the outer world.

Why does expansion cool a gas? Well, if a collection of flying gas molecules is suddenly allowed to expand into a bigger space, the molecules have to push their way out against whatever happens to be occupying that space—usually, the atmosphere. Doing that uses up some of the gas's energy, and the gas molecules then move more slowly. (If the gas is expanding into a vacuum, all bets are off.) A gas whose molecules are moving more slowly is by definition a gas that has a lower temperature.

Did you overshoot while putting air in your tires at the gas station? Do you have to let the excess out? Note that as the excess air comes out of the valve, it is colder than you might have expected. Even on a hot day.

as they can get, within their confines. To force them closer together—to compress them into the confines of a tire, for example—you have to oppose their outward-flying proclivities with some inward-pushing force. When you use your pump, the sweat on your brow tells you that you are indeed putting some of your own muscular energy into that gas. But what do the molecules do with that energy? Unable to fly so far afield anymore, they use the energy you've given them to fly faster. And faster-moving molecules are hotter molecules; heat is nothing but fast-moving molecules. Thus, your muscular energy goes into heating up the gas in the tire.

It doesn't get as hot, however, as you might get under the collar at the idea of having to pay the gas station a dollar for a few pounds of air.

TRY IT

The next time you fly on a humid day, watch the airplane's wing during takeoff, the time of maximum lift. You may see a layer of fog just above the wing's top surface. It's an example of expansion cooling of a gas. The air going over the top of the wing is expanded, compared with the air underneath the wing. (Bernoulli's principle and all that; ask any pilot.) The expanded wing-top air can be cooled enough to condense water vapor out of the air, leaving a stream of visible fog.

A Pair of Highly Extinctive Gases

Carbon monoxide and carbon dioxide: what's the difference? I gather that monoxide means one "oxide" and dioxide means two of them. That's all right with me, but are they both poisonous? What is their connection with automobile exhausts, kerosene heaters, and cigarette smoke?

They're both dangerous gases, but in very different ways.

Small amounts of carbon dioxide are normally present in the atmosphere. It gets there from volcanos, from the decomposition of plant and animal matter, from the burning of coal and petroleum, and from the opening of cans of beer, which however is not the primary source in spite of the way it looks in television commercials. Nevertheless, 11 billion pounds of carbon dioxide are produced annually in the US alone, and much of it destined to be burped into the atmosphere via the eight billion cases of carbonated soft drinks and 180 million barrels of beer that Americans guzzle each year.

Obviously, carbon dioxide cannot be toxic in itself. The only real problem is that it doesn't support burning or breathing, and if given the opportunity, it will extinguish both fires and people. Because carbon dioxide is heavier than air, it will spill down to the lowest level and hang around like an invisible blanket, replacing the air and suffocating anything it covers. That's what happened in Cameroon, Africa, in 1986, when Lake Nyos belched an enormous, 600-ton bubble of volcanic carbon dioxide gas that spread out over the countryside. It suffocated more than 1,700 people and innumerable animals.

Carbon monoxide (CO), on the other hand, is a real villain, even in tiny amounts. When breathed, it goes straight from the lungs into the blood stream, where it reacts vigorously with the hemoglobin, preventing it from doing its vital job of carrying oxygen to the cells. Oxygen deprivation ultimately leads to a condition known as death. Carbon monoxide is the principal cause of poisoning fatalities in the US.

Whenever carbon-containing substances burn in air—from the gasoline in a car to the kerosene in a heater to the tobacco in a cigarette—carbon monoxide is formed to some extent. If they had an unlimited supply of air, these fuels would burn completely, all the way to carbon dioxide—two oxygen atoms in each molecule. But there is always a practical limit to how fast the oxygen can feed itself into the conflagration. So invariably, some of the carbon atoms will manage only to latch onto one oxygen atom instead of two. Result: monoxide instead of dioxide.

Automobile engines spew out about 150 million tons of carbon monoxide in the US each year. In a traffic jam, the carbon monoxide level in the air can build up to sickening (fatigue, headache, nausea), if not dangerous, levels. Kerosene heaters, gas, space, and water heaters, gas furnaces, gas ranges and ovens, gas dryers, wood stoves, charcoal grills, and cigarettes all produce carbon monoxide, and all must be vented to the outdoors or used in a well-ventilated environment.

So don't smoke or drive. Especially indoors, when the kerosene heater is on.

TRY IT

Light a votive candle—a candle in a small glass cup. Don't bother to pray. Now make some carbon dioxide by pouring a little vinegar (acetic acid) onto a few teaspoons of baking soda (sodium bi-carbonate) in a tall drinking glass. As the chemi-cals react, they produce carbon dioxide (CO_2) gas, which bubbles up and fills the glass. Lift the glass and pour over the votive candle as if you were pouring an invisible liquid. (Be careful not to pour any of the real liquid.) The candle will go out, drowned beneath a sea of unseen gas.

A Thousand Pounds of Pigeon Sweat

At a truck stop, I watched a trailer-truck driver banging fiercely on the sides of his trailer with a baseball bat. When I asked him what he was doing, he explained, "My rig is a thousand pounds overweight. I'm hauling 2,000 pounds of pigeons, and I've got to keep half of them in the air at all times." Okay, so it's a joke, but would that really work?

A very old joke indeed, but with an intriguing scientific hook.

No, it wouldn't work.

Think of it this way. The trailer is a box full of stuff. The box weighs so many pounds. Can banging on it possibly change its weight, whether it happens to be filled with gold bricks, sand, goose feathers, pigeons, or butterflies? Obviously, not. The weight of a collection of material is the sum-total of the weights of the molecules in it, no matter how you rearrange them.

But what throws many people is the fact that airborne butterflies and pigeons are not resting on the floor, like other kinds of cargo. So how can their weight be transmitted to a scale that an inspector might place beneath the truck?

Through the air.

Air is, after all, a substance, albeit a thin and invisible one. It is made of molecules like everything else, and it therefore has weight: 1.16 ounces per cubic foot at sea level, to be specific. The terrified pigeon who is catapulted into unplanned flight stays up in the air by repeatedly pushing down upon the air with its wings. (This is an oversimplification of bird flight, but it will do.) When the wing presses down upon the air, the push is transmitted, molecule by molecule, throughout the air. (You would be able to feel the breeze if you were there, wouldn't you?) The pressed-upon air in turn presses upon everything it is in contact with, including the walls, floor and ceiling of the trailer. The pigeon's wing-push force therefore remains completely within the trailer and doesn't change its effect upon a scale.

But, you might say, when the pigeon takes off, doesn't it push down on the floor of the trailer, making it instantaneously heavier, instead of lighter? And even after the pigeon is in the air, don't its downward wing-thrusts constitute an extra downward force on the trailer via the air, again making it instantaneously heavier?

Right on both counts. But according to Sir Isaac Newton, a man who knew what he was talking about if ever there was one, every reaction has an equal and opposite reaction. Thus the downward push on the trailer is exactly cancelled by an equal upward push on the pigeon. Come to think of it, that's why it flaps its wings in the first place.

Perhaps what the truck driver should have done was to install a drain in the floor, introduce a cat into the trailer, and drain out the pigeon sweat as it accumulated.

Nitpicker's Corner. No, pigeons don't sweat.

CHAPTER 5

The Great Outdoors

Step outside with me, please. Look beyond everything that has felt the hand of humankind. Look at the air, the sun in the sky, the clouds. And marvel at it all.

How can something as insubstantial as air exert a pressure on us? Why does the sun feel hotter at certain times of day? Why are some clouds black? Why does it get warmer when it snows? And if you have ever been to a beach, haven't you wondered why the waves always roll in in the same way, no matter whether the coastline faces north, east, south, or west?

To paraphrase Charles Dudley Warner (and no, it was not Mark Twain), nobody does anything about the weather, but we sure can talk about it. Even better than talking about it is understanding it, by observing it closely and thinking things through. In this chapter, we'll experience sunshine, clouds, wind, and snow. And along the way we'll comment on a couple of human-made outdoor phenomena: the Statue of Liberty and Fourth-of-July fireworks.

By the Beautiful Sea

Every time I'm at the seashore, there seems to be a cool breeze blowing in from the ocean. Is it my imagination, or is there something about the shore that makes it inherently cooler and windier?

You're right. "Sea Breeze" isn't just the name of a thousand beach motels. The breeze coming in from the sea is a real phenomenon that makes the shore cooler than it is inland—at least in the afternoon, which is when people most want to cool off anyway. In the daytime, cool breezes almost invariably blow in from the ocean toward the land, rather than the other way around. They begin several hours after sunrise, reach their peak by midafternoon, and die out toward evening.

What happens is that, starting in the morning, the sun beats down on both land and sea. But the sea is not noticeably warmed by the sunshine because it is so cold and vast that it has an inexhaustible appetite for heat energy, slurping it up with nary a degree's rise in temperature. The land, on the other hand, is substantially warmed up by the sun's rays; soil, plant leaves, buildings, roads, and so on are relatively easy to heat. (They have low *heat capacities,* compared with water.) As the land warms up, it warms the air above it, which expands and rises. The cooler, denser air that is sitting over the water then rolls in underneath it, sweeping over the beach and cooling the bodies of bikinied beauties.

It's not just that the ocean breeze is cool. Even if it weren't, it would still be helping to cool the sweltering hordes by evaporating perspiration from their extravagantly exposed epidermises.

Doin' the Wave

At the seashore, why do the waves always break in lines parallel to the shore, regardless of the direction in which the shoreline runs?

Waves can tell when they're approaching a shore, and they actually turn to line up with it.

What makes waves, of course, is wind blowing across the water's surface. But it can't be that the wind is always blowing the waves straight in to shore. Out in the ocean, the wind may be blowing every which way. The waves we observe at the shore are only those that are traveling more-or-less generally in our direction, or else we'd never see them. Nevertheless, most of them approach obliquely, not perpendicular to the shoreline. What happens then, believe it or not, is that the incoming wave "feels" the shore and turns to face it squarely before breaking. Later, when it breaks (see below), the line of foam will be quite parallel to the shoreline.

The question, of course, is how a wave knows when it is approaching a shore. And what makes it turn?

When a wave—consider it a broad bump on the surface—is still over deep water, there's nothing to restrain it; it goes wherever the wind commands. But as it moves into shallower water, the lower part of the wave begins to drag on the bottom, which slows it down. That's its clue that it is approaching a shore, and that's what gives it a preferred direction.

Let's say we have a wave that is coming in at an angle, with the shoreline to its left. The first part of the wave to hit shallow water and scrape bottom will be the left end of the wave. The left end will therefore be slowed down, while the middle and right end keep going at the same speed. This has the effect of turning the wave to the left—toward the shore. (If you drag your left foot from a go-cart, you're going to swerve to the left, aren't you?) This dragging and slowing proceeds down the wave's length from left to right as more and more of the wave feels the drag, and gradually the whole wave is turned to the left. Its crest line is now stretched out parallel to the shoreline, and that's the position it finds itself in when it gets close enough to the shore to break.

Waves break because of the same bottom-dragging effect. After the wave has lined up parallel to the shore, it eventually reaches such shallow water, and its bottom is slowed down so much, that its top overtakes its bottom and tumbles over it. The top falls with a crash, churning up a line of foam all along the wave's length. And that's parallel with the shoreline.

TRY IT   *The next time you fly over a curving coastline, notice how the white lines of foam from the breaking waves are always parallel to the shore, no matter which way the shoreline turns.*

Ever on Sun Days

Why do they say that the risk of sunburn is greatest between the hours of 10 a.m. and 2 p.m.? Of course, that's when the sun is most directly overhead, but why is the overhead sun stronger? It's not any closer to us at noon, is it?

No, the 93-million-mile separation between the sun and the Earth pays little attention to our lunchtime or recreational schedule. The sun is essentially the same distance from your rapidly reddening nose at all times of day. But the strength of the sunshine varies, for two reasons: an atmospheric reason and a geometric reason.

Picture the Earth as a sphere, covered by a layer of air—the atmosphere—a couple of hundred miles thick. When the sun is directly overhead, its rays are coming down perpendicular to the atmosphere

and to the ground, penetrating the least possible amount of atmosphere in the process. But when the sun is lower in the sky, its rays are coming to us obliquely and somewhat horizontally, having to penetrate much more of the atmosphere before getting to us. Because the atmosphere scatters and absorbs some sunlight, the more atmosphere the rays have to penetrate, the less intense they become. So low sun is weaker in intensity than high sun. Near sunrise or sunset, it is almost 300 times dimmer than at noon.

But even if the Earth had no atmosphere, the sunlight would still be weaker when the sun is lower in the sky. It's a purely geometric effect of the obliqueness of the rays. The best way to see this effect is with a flashlight and an orange.

Nitpicker's Corner. If we wanted to, we could call this geometric effect the "cosine effect." If you work out the trigonometry, it turns out that the intensity of sunlight on the ground falls off according to the cosine of the angle between straight overhead and the sun's position. The intensity (and the cosine) decreases from full value at high noon on the equator to zero when the sun hits the horizon at sunset.

TRY IT

In a darkened room, shine the round beam of a penlight or tiny flashlight onto the surface of an orange. The penlight is the sun and the orange is the Earth. First, hold the penlight directly above the equator: in noonday position. You'll see a perfectly circular beam of sunlight landing on the Earth. Now, holding the sun the same distance from the Earth (makes you feel powerful, doesn't it?), shine the beam onto the Earth obliquely, a little to the left (west) of where you had it before: in late afternoon position. You'll see an oval-shaped light on the orange, as if the circle of sunlight has been smeared out. Well, it has been. The same amount of light is now spread out over a larger area, so of course its intensity at any one spot must be lower.

The next time you're at the beach, notice that the black-belt sun tanners use this effect to their advantage (and that of their dermatologists' bank accounts). At any time of day, lying down makes the sunshine strike you at a somewhat oblique angle, because it is never directly overhead, except at the equator. What the Olympic tanners do is face the sun and sit up slightly, so that the cancer beams strike their skins as perpendicularly as possible.

You didn't ask, but . . .
Isn't that why it is colder in the winter than in the summer?

Right on. When it is winter on the part of the Earth where you live (northern or southern hemisphere), your hemisphere is leaning away from the sun a bit. That is, the axis of the Earth wobbles, so that during winter in the northern hemisphere the North Pole is farther from the sun than the South Pole is. Because your hemisphere is leaning away from the sun, the sunshine hits its surface at a more oblique angle. The more oblique the angle, the less intense the light. And, of course, the heat. Big surprise conclusion: you're less likely to get either sunburn or heat stroke in the winter.

Mad Dogs and Englishmen

In the summertime, whenever somebody wants to impress me with how hot it is, they'll say something like, "It's ninety degrees in the shade." But I can't always stay in the shade. I want to know how hot it is out in the sun, too. Is there any way to translate in-the-shade temperatures to in-the-sun temperatures?

Afraid not. While the temperature "in the shade" is a fairly reproducible figure, the temperature "in the sun" depends too much on whose temperature you're talking about.

Different objects, including different people in different clothing, will experience different temperatures in the sun because they will absorb different amounts of different portions of the sunlight's spectrum. Light-colored clothing, in general, absorbs less—reflects more— of the sun's radiations than dark clothing does, so it keeps us cooler. It's much the same with human skins: a light-skinned person may not feel as hot in the sun as a dark-skinned person would. When British imperialism was at its peak in parts of the world where the people have generally darker skins, Noël Coward immortalized that fact in his song, "Mad Dogs and Englishmen [Go Out in the Mid-day Sun]."

In the shade—in the absence of direct radiation from the sun—the temperature of a free object (not connected to a source or absorber of heat) depends mainly on the temperature of the surrounding air. That's the temperature that the weather people quote in their reports; they don't bother to say "in the shade." But in the sun, temperatures depend not only on the air's temperature but on the absorption and reflection of heat rays by the object or person in question. These factors can vary a great deal from object to object and condition to condition.

And no, there is no physical law that says that steering wheels get hotter than anything else when you park your car in the sun. It's just that the steering wheel is in a particularly sun-vulnerable position and it's the object you need to touch most.

Green Skin and Blue Blood

Those bluish-green roofs on old churches and city halls: I under-
stand that they're made of copper, but I have never seen copper
turn that color anywhere else. Could I make a penny green like
that?

Those copper roofs have been out in the weather longer than a
borrowed lawn mower—all those years since people could afford to
cover roofs with that durable, beautiful red metal. Today, copper is
too expensive to use for keeping rain off the heads of politicians and
priests. It is even too expensive to make pennies out of. Since 1982,
pennies have been 97.5 percent zinc, with just a thin plating of copper
for old times' sake.

But if you really want to, you can still leave a penny out in the
weather for 50 years or so and it will turn roof-green. There's no fast
and easy way to do it.

That's the reason, in fact, that copper is such a good material for
covering roofs: it corrodes very slowly—much more slowly than iron
rusts. Within a few weeks exposed to the air, bright, shiny copper will
begin to darken because of a thin layer of black copper oxide. Then,
as years go by, it reacts slowly with oxygen, water vapor, and carbon
dioxide in the air to form the bluish-green patina that chemists identify
as *basic copper carbonate*.

In addition to roofs, this patina is what colors the Statue of Liberty, which is made of 300 thick copper plates bolted together, and
which has been exposed to New York City air since 1886. Here's
the predominant chemical reaction of copper with New York's atmospheric water, carbon dioxide, and oxygen: $2\ Cu + H_2O + CO_2 +$
$O_2 \rightarrow Cu(OH)_2 + CuCO_3$ to produce a mixture of copper hydroxide
and copper carbonate.

Incidentally, the green color that you see on pennies in the bottoms
of fountains, tossed in by people who believe that one cent will bribe
the Fates into granting a wish and that it will somehow be overlooked
by the midnight scavengers, is not the same, chemically, as the roofs'.
It is due to other compounds of copper such as chloride, hydroxide,
and what chemists call "God-knows-what" that don't have the same
blue-green shade and don't adhere very well to the metal.

You can try to duplicate the patina of copper by buying some cheap
jewelry made of brass, which is an alloy of copper and zinc. Wear an
unlacquered brass ring or bracelet for a few weeks and the copper will

You didn't ask, but . . .

What about those copper bracelets that are supposed to cure arthritis?

Baloney. The thinking (a generous term for it) behind these voodoo baubles appears to be (1) that copper is a good conductor of electricity (which it is), (2) that there is electrical energy in the air (whatever that means), and (3) that a copper bracelet will therefore attract that "energy" and conduct it to your aching bones. And of course, we all know that energy is good for us, don't we?

The only energy the bracelet will generate, however, is the energy you will have to expend in scrubbing the green stain off your wrist.

react with salt and acids in your skin to produce copper chloride and other compounds that will stain your skin as green as Miss Liberty's. But it still won't be the same shade as hers unless you stand out in the New York City air for a hundred years or so.

Many outdoor statues in public places are made of bronze, which is an alloy of mostly copper and tin. When the statues weather, they develop a dark-green patina similar to copper's. The white splotches on the statues have quite a different origin, however.

An interesting sidelight on copper is that instead of the red hemoglobin in human blood, which has an iron atom in its molecule, lobsters and other large crustaceans have blue blood containing hemocyanin, which is similar to hemoglobin but contains a copper atom in place of the iron. There may be some truth, after all, in the claim of revolutionaries that the world's blue bloods are among the lowest forms of life.

It's a Good Thing Air Is Transparent

How come we can see through air?

It's very simple. The molecules in air are so far apart that we're actually looking through empty space. To notice anything at all, we'd have to be able to see the individual molecules, but air molecules are a couple of thousand times smaller than anything we can see with an optical microscope.

We're talking about looking through pure, unpolluted air, of course. We'll get to the dirty stuff later.

Air is 99 percent nitrogen and oxygen molecules, which are roughly equal in size. The figure shows them, drawn to scale, at their normal separation distance at sea level. Notice all the absolutely empty space: nothing whatsoever in between the molecules. No wonder light can pass through air directly to our eyes, completely unhindered. And that's as good a definition of transparency as any.

But even when visible light happens to hit one of the nitrogen or oxygen molecules, it isn't absorbed. Many other kinds of molecules have the habit of absorbing light of certain specific wavelengths, or colors, or energies. When a certain specific color is absorbed out of the light, the rest of the light, lacking that color, appears to us as an

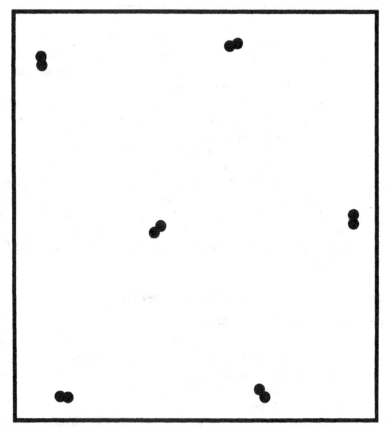

A scale drawing of the air at sea level.

altered color. So some gases appear colored, even though we can still see through them. Chlorine gas, for example, is yellowish green. If you had a glass jar full of chlorine gas, you would still be able to see through it because the molecules are still very far apart, but the light coming to your eyes would have a greenish tinge.

Transparency and color, then, are two different things, in spite of the fact that many people call colorless plastic "clear" instead of color-less. Tinted glass may be colored, but you can still see through it; it is still transparent.

Which brings us to polluted air: air you can see. If you have ever flown into Los Angeles, Denver, or Mexico City, you may have seen a layer of yellowish brown sludge hanging over the city. That is air con-taining nitric oxide, an irritating gas that is made when other nitrogen oxides from automobile exhaust react with oxygen in the air.

When pollutants, including smoke and chemical fogs, get so thick that a variety of light wavelengths are being absorbed, the air be-comes generally less transparent. The molecules are still far apart, but so many of them are absorbing or scattering light that less of it gets through to our eyes. There are many locations in the "wide open spaces" where the visibility has diminished so much over a single life-time that adults can no longer see the distant mountain peaks that they could see clearly when they were kids.

Yes, we're lucky that air is transparent. But it isn't as transparent as it used to be.

Living Under Pressure

Why do weather people talk about the barometric pressure in terms of so many inches of mercury? How can you measure pres-sure in inches? And what is an inch of mercury, anyway? It's a liquid, isn't it?

The phrase barometric pressure is an odd locution. The atmosphere around us has a temperature that is measured by a thermometer, a humidity that is measured by a hygrometer, and a pressure that is mea-sured by a barometer. Television weather reporters would not dream of talking about the air's "thermometric temperature" or "hygroscopic humidity," yet they insist on referring to the air's "barometric pres-sure," probably because it makes them sound impressively scientific. Plain old "air pressure" or "atmospheric pressure" would do just fine.

But what is this "pressure" that the air exerts? The atmospheric pressure is caused by the incessant bombardment of air molecules upon whatever they happen to be in contact with. Every time an air molecule (primarily nitrogen or oxygen) crashes into a solid or liquid surface, it exerts a force. The number of these collision forces per second on each square inch or square centimeter of the surface is a measure of the pressure. And it is not trivial; at sea level, those zillions of molecule collisions add up to 14.7 pounds on every square inch (1.03 kilograms per square centimeter).

Measuring pressure by counting molecular collisions is difficult. But since the atmosphere exerts a pressure on everything it is in contact with, we can use the total force upon any convenient pressed-upon object as our measuring standard.

In 1643 in Florence, Italy, Evangelista Torricelli decided that the atmosphere's pressure should be capable of forcing water up into a tube to a certain height, and that the height of that column of water would be a measure of the atmosphere's pressure. He thus invented the world's first barometer. It turned out that normal atmospheric pressure would support about 34 feet (about ten meters) of water. But that would make for some ridiculously large barometers. So today we use the much heavier liquid, mercury—a silvery liquid metal—which on a typical day by the seaside can be forced only 29.92 inches or 760 millimeters up a tube.

Another way of saying this is that the atmosphere is applying the same amount of pressure that we would feel if we were submerged under 34 feet of water or under 29.92 inches in a pool of mercury.

TRY IT

To get an idea of how much pressure the atmosphere is exerting on us, stick your toes under the leg of a kitchen or dining room chair and put a 10-pound sack of potatoes on the seat. That will apply a pressure of roughly 10 or 15 pounds per square inch to your toes. That's in addition, of course, to the 14.7 pounds of pressure from the atmosphere, but you don't feel that because it is uniform all over your body. Do fish know they're under water? We're under air.

Look Not for a Silver Lining

Why are clouds white, except when they're storm clouds, and then they're black?

It's all a matter of how big the water droplets are.

That's what clouds are: tiny droplets of water. They are so small that under the continual bombardment of air molecules they are kept suspended in the air and don't settle out by gravity—until it rains, of course. The droplets do keep evaporating and re-forming, however, and that's why clouds keep changing their shapes.

The droplets of water in a white cloud are like tiny crystal balls. That is, they reflect and scatter light in all directions. Like water in other forms, they reflect and scatter all wavelengths (colors) of light equally, so the reflected sunlight reaching us retains its full white color. When the droplets are even smaller, smaller than the wavelengths of light, they contribute to the blue color of the sky.

TRY IT

On a day when there are white clouds moving through a clear blue sky, lie down and watch them for a while. You'll see that they continually change their shapes as they move along with the wind. The water droplets at the edges are continually evaporating and re-condensing elsewhere, which changes the outline of the cloud.

On the other hand, storm clouds, as you might expect, are loaded with water, just waiting for the right opportunity to spoil your picnic. The droplets of water in them are so thick that they block out light coming down from the sun and the cloud appears dark against the sky. They're not actually black, however, any more than a shadow is.

But if you're looking for a silver lining, check out Liberace's or Elvis's costumes.

Not Quite Cricket

I read somewhere that you can tell the temperature by listening to the crickets. How?

You count their chirps.

All cold-blooded animals perform their functions faster at higher temperatures. Just watch how fast the ants run around in cool and hot weather. Crickets are no exception; they chirp at a rate that is geared directly to the temperature. To understand their message, all you need is the translation formula.

This is not so much a biological phenomenon as it is a chemical one. All living things are governed by chemical reactions, and chemical reactions generally go faster at higher temperatures. That's because chemicals can't react with one another unless they come into contact: molecules actually bumping into molecules. The higher the temperature, the faster the molecules are moving and the more often they will collide and react. Chemists like to use the rule of thumb that a chemical reaction doubles in speed for every 10-degree (Celsius) rise in temperature.

Fortunately, we warm-blooded critters maintain a constant temperature and therefore a pretty constant rate of living. Crickets, however, chirp faster when they're warmer. The best one to listen to is the snowy tree cricket of North America. But if you can't tell one cricket from another, don't worry about it. The common field cricket chirps at about the same rate.

Here's how to tell the temperature by listening to a cricket: Count the number of chirps in 15 seconds and add 40. That will give you the temperature in °F.

When the United States finally switches to the metric system, crickets will be required by law to chirp in Celsius. You will then be able to determine the temperature in °C by counting the number of chirps in eight seconds and adding five.

Be aware that the cricket is broadcasting the temperature where it happens to be. Unless you're up a tree or in the grass, your temperature is not quite the cricket's.

People Who Live on Glass Planets Shouldn't Burn Coal

When I entered the greenhouse at a plant nursery, I was struck by how much warmer it was than outside. Is it always warmer in a greenhouse? If so, why?

Yes, greenhouses—sometimes called hothouses or glasshouses—are always naturally warmer, without any artificial heating. But believe it or not, the main reason is not what everybody refers to as "the greenhouse effect."

A greenhouse is just a closed, glass container for plants. The glass lets in sunlight, which the plants need for growth, while keeping out damaging wind, hail, and animals. It also prevents the loss of moisture and keeps the humidity high, which is part of what hit you in the face when you entered. But the main thing it does is to act as a heat valve, limiting the loss of heat from the plants to the cold, cruel world outside.

A plant, or anything else for that matter, can get colder—that is, can lose heat—in any of three ways: by *conduction*, by *convection*, and by *radiation*. Conduction isn't a problem because the leaves aren't in contact with anything, such as a mass of metal, that could conduct their heat away. That leaves convection and radiation. The greenhouse cuts down both.

Convection is the circulation of warm air. Because warm air rises, it can carry heat up and away from a plant leaf. Anything that prevents that warm air from escaping completely will prevent the loss of heat along with it, and any closed-in building will serve that purpose. That's the major effect of the greenhouse: it simply prevents heat loss by breezes and currents of air. Of course, no farmer would dream of constructing a plant-enclosure building without letting in lots of sunlight, and that's how glass-walled and glass-ceilinged greenhouses were born.

A secondary effect of the glass, which nobody knew when they invented greenhouses, is that it cuts down heat loss by radiation. That's where the so-called greenhouse effect comes in, and here's how it works.

The photosynthesis reactions that keep plants living and growing use ultraviolet radiation from sunlight. After using some of the energy

You didn't ask, but . . .

How is the greenhouse effect connected to global warming?

The Earth's atmosphere can trap infrared radiation just as the greenhouse's glass does. And that can raise the average temperature beneath it—namely, at the surface of the entire globe.

The overall temperature of the Earth's surface averaged over all seasons and climates depends on a fine balance between the amount of the sun's radiation that comes down to us and the amount that is reflected or reradiated out into space. About one-third of the sun's energy that hits the Earth is reflected back out; the rest is absorbed by the clouds, land, sea, and John Boehner. Most of the absorbed energy soon degenerates into heat, or infrared, radiation, just as it does in the plants in the greenhouse.

Hanging over the Earth's radiating surface is a transparent canopy similar to the glass in the roof of a greenhouse. It's a layer of air: the atmosphere. Like glass, air is quite transparent to most of the sun's incoming radiations. But certain gases in the atmosphere, mainly carbon dioxide and water vapor, are very efficient absorbers of infrared radiation. Just as the glass does in the greenhouse, these gases block the escape of some of the infrared, trapping it down here at the Earth's surface, which is therefore somewhat warmer than it would be if there weren't any carbon dioxide and water vapor in the atmosphere.

But recent human activities have been changing the balance between infrared coming down from the sun and infrared being radiated back out through the atmosphere. Ever since the industrial revolution began about

---

of this radiation, they emit lower-energy "waste radiation": infrared radiation, which can then be absorbed by other objects. But when an object absorbs infrared radiation, it grows warmer. We can therefore think of the infrared radiation from the plants as if it were heat, traveling through the air in search of something to warm up.

What happens when the infrared radiation hits a glass wall or ceiling? Although glass lets in ultraviolet light pretty well, it is not completely transparent to infrared. So the glass blocks some of the infrared radiation from getting out of the greenhouse, and this trapped radiation gradually warms up everything inside.

Clearly, this heating can't go on forever; greenhouses have not been known to suffer spontaneous meltdowns. After a certain point,

a hundred years ago, we have been burning coal, natural gas, and petroleum at an ever-increasing rate. When these fossil fuels are burned, they put carbon dioxide into the air. The amount of carbon dioxide in the atmosphere has therefore risen substantially in the past century. More carbon dioxide means more earth-trapped infrared radiation and higher temperatures.

The amount of global warming that can be caused by a given amount of carbon dioxide in the atmosphere is not easy to determine. On the one hand, the oceans and forests diminish the effect by soaking up carbon dioxide from the air. On the other hand, the world's huge rain forests are rapidly being devoured by logging and by burning, which compounds the problem by putting even more carbon dioxide into the air. But it has been well established that the Earth's average temperature has been rising, and that it may rise by another 1.5°C to 4.5°C (0.8°F to 2.5°F) if we continue burning fossil fuels at the current rate.

A temperature increase of only a few degrees could have catastrophic consequences. Slightly warmer Arctic and Antarctic climates would melt huge amounts of ice, raising the level of the oceans and inundating coastal cities all over the world. Also, there would be changes in global weather patterns, with significant consequences in food production and water supply.

Our planet's atmospheric greenhouse is apparently just as fragile as if it were actually made of glass, and like vandals, we keep throwing carbon dioxide at it instead of rocks.

the inevitable leakage of heat out of the house balances the infrared buildup inside, and the temperature levels off at a moderately warm level, warmer than if the glass were completely transparent to infrared radiation.

Ridiculous? No, Sublime

When there is snow on the ground, I've noticed that it slowly melts away over a period of a week or two, even when the temperature stays well below freezing. Where does it go?

The snow isn't melting; it is actually evaporating. It's going straight off into the air as water vapor, without having to melt into liquid water first. But scientists prefer to reserve the word "evaporation" for liquids only, so when a solid evaporates, they call it *sublimation*.

Nobody is surprised when water evaporates—when it changes from liquid to vapor. But in our everyday experience, we rarely notice the evaporation of solids, claims to the contrary notwithstanding. ("I don't know where all the money went. I swear, it just evaporated.") Solids generally evaporate too minutely for us to notice. Nevertheless, sublimation has probably played at least two roles in your life: as moth balls and as freeze-dried coffee. We'll get to them shortly.

The molecules at the surface of a solid aren't tied down as firmly as the molecules are within the body of the piece. The molecules in the body are bonded to their brethren in all directions, top, bottom, and all around; the surface molecules, on the other hand, are bonded in every direction but their "tops", which are exposed to the great outdoors. They are missing a bit of adhesion to the rest of the solid. If you consider that molecules are always jiggling around to some extent, it is not too hard to imagine that an occasional surface molecule might enjoy such a jolly jiggle that it breaks loose and flies off into the air. That molecule has evaporated, or sublimed.

The molecules of liquids are more loosely tied together than the molecules of solids are, so the probability that a liquid molecule will break away is much greater. That's why liquids evaporate much faster than solids do.

Snow is a great candidate for sublimation because it is made of intricate, lacy crystals with large surface areas, and the more surface molecules there are, the more molecules can evaporate. But you can even see solid chunks of ice sublime. Ever notice how old ice cubes shrink in the freezer? Different solids have different tendencies to *sublime*—that's the verb, not "sublimate"—because they are made of different atoms or molecules that are tied together with different strengths. Fortunately, the atoms of metals are tied together very tightly, so gold and silver don't evaporate at all. On the other hand, the molecules of some organic solids are tied together rather loosely, so they have a substantial

---

You didn't ask, but . . .
How is freeze-dried coffee made?

Now why would I put a question like that in a chapter on The Great Outdoors? But wait; it'll all come clear.

Freeze-dried coffee is made by *sublimation*, the process by which a solid, such as ice, for example, changes directly into a vapor without having to melt or boil first.

To make either instant or freeze-dried coffee, they first brew 2,000-pound batches of incredibly strong coffee that could keep China awake for a month. From then on, freeze-dried and instant coffees go their separate ways.

If they are making instant coffee, they quick-dry this thick brew by waterfalling it down through a high-temperature chamber that evaporates all the water, and only the powdered solids fall to the bottom. Unfortunately, though, the heat drives off some of the most flavorful and aromatic chemicals (over 800 different chemicals have thus far been identified in coffee aroma), so instant coffee has never been the darling of connoisseurs.

On the other hand, when making freeze-dried coffee they freeze the strong brew into blocks of what could be called Popsicle coffee. Then they pulverize them into granules and put them in a vacuum chamber, where the water flashes off by sublimation—no heat involved—leaving all the flavor chemicals intact. Most connoisseurs of coffee-in-a-hurry believe that, compared with ordinary instant coffee, freeze-dried coffee tastes sublime.

tendency to fly off as vapor. Moth crystals and deodorizing cakes are usually made of naphthalene, an organic solid that is a sublime evaporator, so to speak. Its strong-smelling vapor quickly fills the air and kills both moths and our ability to smell foul odors.

Sizzlin' Snowflakes, Batman, Why Is It Getting So Warm Around Here?

This is going to sound crazy, but I swear it's true. I spend a lot of time outdoors in the winter, and every time it begins to snow, I've noticed that the air becomes warmer! You'd think that in order to start snowing, the air has to get colder, not warmer. What's going on?

You're not crazy. You're a good observer. It really does get warmer when it begins to snow.

Think of it this way: In order to *melt* a lot of ice or snow, you have to add heat to it. So when a lot of water *freezes* into ice or snow—the reverse process—that same amount heat has to come back out again. It does, and it heats up the air. The question is why that heat comes out.

First of all, water in the air isn't going to freeze into snowflakes at all unless the temperature is lower than 32°F or 0°C. You've never heard a weatherman predict "temperatures in the mid-seventies with occasional snow flurries," have you? So all the necessary cooling—or temperature lowering—that you rightfully expect will already have taken place by the time the first flake forms. Nothing the least bit remarkable there.

You didn't ask, but . . .
When frost threatens, why do people spray their tomato plants with water to protect them from freezing?

The water on the wet plant leaves are exposed to the air and will begin to freeze first, releasing its 80 calories of heat per kilogram. The leaves absorb this heat and stay warmer than they would otherwise have been. Gardening books are wrong when they tell you that the frozen water protects the leaves by acting as an insulator. The insulation value of a thin coating of ice is nil.

Bar Bet. A snowfall makes the weather warmer.

As soon as the water begins to freeze into snow, though, something new begins to happen. In a droplet of liquid water, the molecules are quite loose; they're sliding around each other freely and randomly. But when that droplet freezes into the beautifully shaped ice crystal we call a snowflake, the water molecules must snap into a rigid crystalline formation. They have less energy in the rigid snowflake formation than they had in the chaotic liquid form. It's like a schoolteacher taming a bunch of wild kids by making them line up in the hallway. If the water molecules now have less energy in the snowflake than they had in the water droplet, the excess energy had to go somewhere. It did. It went off into the air as heat.

For each kilogram of water that freezes into a kilogram of ice or snow, 80 calories of heat are released. If it stayed in that kilogram of water, that amount of heat would be enough to raise its temperature from freezing to 80°C or 176°F! But of course, the heat doesn't stay there, or the water would never freeze. It is immediately swept away into the surrounding cold air.

Thus, when any given kilogram of water turns into snowflakes, the local surrounding air gets a gift of 80 calories of heat. Multiply this by the zillions of kilograms of water that are freezing at the beginning of a snowfall, and it's no wonder you feel warmer.

It's All in the Snow-How

As a skier, I often have to settle for artificial snow, made by those snowmaking machines. Do they just pump a spray of water into the air and let it freeze?

No. That wouldn't work very well, except perhaps in extremely cold weather. And by the way, the machines don't produce actual snow-flakes; they make tiny beads of ice, around ten thousandths of an inch in diameter.

The simple spraying of water wouldn't work because when water freezes, it gives off quite a bit of heat. That happens because when water molecules transform themselves from a liquid to a solid, they have to stop moving around and settle down into rigid positions, and the movement energy that they previously had simply must wind up someplace. If large amounts of sprayed water were to be frozen fairly near the ground, the liberated heat would warm up the air substantially and defeat much of the whole purpose; the ersatz snow would be wet and not very cold. When real snow forms in nature, on the other hand, the heat is given off way up in the air someplace where the snow was created, and it doesn't significantly warm up those beloved slopes. That's why at many ski resorts the snowmaking machines do their spraying from high towers, letting the wind carry off the heat.

In any event, some extra cooling is necessary to counteract the heat that is released on freezing. The machines accomplish this by spraying not just water, but also a mixture of water and high-pressure air, at around 118 pounds per square inch. When compressed air, or any gas for that matter, is allowed to expand suddenly, it gets cold, because in pushing aside the atmosphere or anything else that it's expanding against, it uses up some of its energy. The coldness of the expanding air more than makes up for the warming that comes from the freezing water. And then for good measure, the blasted-out water droplets are cooled further by evaporation.

Strangely, though, no matter how cold the water gets, it won't spontaneously freeze. People say that water freezes at 32°F (0°C), but they should add, "provided that something stimulates it to begin the freezing process." Water molecules can't begin to settle into their highly specific orientations and rigid positions that they must have in an ice crystal without some kind of "starting gun" to shake them into place. It's a fact that water can be cooled far below the normal freezing temperature—that is, it can be substantially *supercooled*—without freezing. This is much too tricky to try at home, but under careful laboratory conditions, pure water can be supercooled down to –40 degrees without freezing. (Fahrenheit or Celsius, it doesn't matter; 40 below happens to be 40 below on both scales.

A mechanical shock can shake up the molecules in supercooled water enough to make them fall into their assigned places in an ice crystal. In the case of the snowmaking machine, the shock is supplied

TRY IT   *Measure the length of a convenient icicle during a cold spell. Then come back in a few days and measure it again. Make certain that the temperature hasn't gotten above freezing in the meantime, so that there hasn't been any melting. You will see that the icicle has gotten smaller by sublimation.*

by the high-pressure air blast, which shoots the microscopic droplets of water out of nozzles at near-sonic speeds. An interesting new wrinkle in snowmaking machines is the addition of a certain species of harmless bacteria, *Pseudomonas syringae*, which live on plant leaves almost everywhere, to the sprayed water/air mixture. Proteins in these bacteria serve as *crystallization nuclei* that initiate the formation of ice crystals at relatively high temperatures, so that the droplets will turn into ice before evaporating or falling to the ground.

Snowball Fight!

I've been having an argument with a friend about what holds snowballs together. He said that because snowflakes are jagged, they must hook together like Velcro. I wasn't convinced. Was he right?

He doesn't stand a snowball's chance. It's a nice idea, because snowflakes certainly do have beautifully complex shapes, with spikes, lacy edges, and all the rest. But interlocking hooks and loops are a bit too much to expect. Besides, they're much too fragile and brittle; when you pack them together, they suffer a crushing experience.

The answer lies in the fact that pressure can melt frozen water: ice or snow. When you press the snow together tightly, the pressure melts certain feathery portions of the flakes. They then can slide over one another on the resulting water film, and the ball compacts. But the main body of the snow is still below the freezing temperature, so the melted parts quickly refreeze. This refrozen ice acts like a cement that holds the whole thing together.

If you are intrepid enough to be making snowballs with your bare hands, your body heat is also melting a thin layer on the outside

---

You didn't ask, but . . .
Does it ever get too cold to make snowballs?

Yes. Every northern kid knows that wet snow makes the best snowballs. That's because snow that's not too much colder than the freezing point is easy to pressure-melt, and it will therefore compact into an effective projectile. But when the snow is too cold, the strength of even the most belligerent bully will be inadequate to pressure-melt and refreeze many flakes, and the snow will fall apart into useless shrapnel.

surface. When this layer refreezes, you've got yourself a case-hardened weapon. Although the Geneva Conventions strictly forbid it, some combatants actually dip their snowballs in water to make them even harder.

TRY IT   *For Yankees only: Put a dark-colored dish in the freezer and wait for snow. When it starts snowing (that's usually when the flakes are biggest), take the dish outside with a magnifying glass, the most powerful one you can find. A cold microscope and slide would be even better. Catch some snowflakes on the dish or slide and quickly examine them with the magnifier. Wow! What beautiful crystals! If the snow catches you unprepared, a piece of cold, dark cloth will also work as a flake catcher.*

Things that Go Boomp in the Night

How do they make all those colors in fireworks?

They add chemicals to the explosive mixtures that emit specific colors of light when subjected to heat. You could throw some of these same chemicals into your fireplace if you thought that a green fire, for example, might be more romantic.

When you throw an atom into a fire, it can pick up (*absorb*) some of the fire's energy and use it to *excite* some of its *electrons*—that is, kick them up to higher *energy states*. These "hot" electrons are just dying to return to their natural energy states, their *ground states*, and the easiest way for them to do that (easy for electrons, that is) is to discharge their excess energy in the form of a *photon* of light. When enough atoms in a fire are simultaneously taking on heat energy and throwing it back off in the form of light, we can observe a bright light show.

Every type of atom has a unique set of possible electron energies, like rungs on a ladder; we say that the energy levels are *quantized*. Each type of atom in the flame will therefore be able to absorb or emit only those amounts of energy—that is, wavelengths or colors of light—that correspond to the "spaces" between its specific set of rungs. That is, each type of atom has its own spectrum of light colors that it can emit after being excited by a flame or explosion.

Unfortunately for the fireworks manufacturers, most atoms emit their light in colors that humans can't see: in the ultraviolet or infrared

You didn't ask, but . . .
How do they make all those colors in neon signs? Is it just colored glass?

No, the colors are actually glowing atoms, stimulated by electricity. It's pretty much the same as making the colors in fireworks: stimulate atoms with energy, and they'll quickly get rid of the excess energy by emitting light of their own characteristic colors.

There are a couple of differences (fortunately) between fireworks and neon signs. In neon signs, the atoms are in the form of gases inside of glass tubes that are shaped to spell out words. Instead of by explosions, the gas atoms are stimulated, or excited, by a high-voltage electric current passing through the tube from one end to the other. If the gas happens to be neon, it emits that familiar orange-red color that announces the presence of Nick's Bar and Grill. Other gases give off their own colors of light when excited by an electric current. For example, helium makes a pink-violet light, argon makes bluish-purple, krypton makes a pale violet, and xenon makes blue-green. Other colors are made by mixing gases or by coating the insides of the tubes with solid materials (phosphors) that glow with their own colors.

Nobody has yet been able stop people from calling all of these signs "neon," regardless of what gas happens to be inside the tubes.

regions of the spectrum. But the atoms of some elements do emit light in brilliant colors that we can see.

Here are some of the kinds of atoms (in the form of their chemical compounds) that are used to make the colors in fireworks: Reds: strontium (used most often) makes a crimson light, calcium makes a yellowish red, and lithium makes carmine. Yellows: sodium makes a bright, pure yellow. Greens: barium (used most often) makes yellowish green, copper makes emerald green, tellurium makes grass green, thallium makes bluegrass green, and zinc makes a whitish green. Blues: copper (used most often) makes azure, while arsenic, lead, and selenium make light blues. Violets: cesium makes bluish purple, potassium makes reddish purple, and rubidium makes violet.

Bar Bet. That blue "neon" beer sign has no neon in it at all.

TRY IT

The next time you have a fire going in your fireplace or on the beach, sprinkle some crushed table salt or some powdered bicarbonate of soda on it and you'll see the brilliant yellow flame color that sodium makes. If you have one of those salt substitutes around—the kind that's intended for people on sodium-free diets—toss some of it onto the fire. It contains potassium chloride instead of sodium chloride, and you'll see potassium's characteristic reddish purple flame color. If you're taking lithium for a bipolar (manic-depressive) disorder, your medicine will make the most beautiful red flames you've ever seen.

Up, Up, and a . . . Why?

What eventually happens to a helium-filled balloon when you let it go outdoors? And why do helium balloons fall up, anyway? Doesn't gravity act on helium, as well as on everything else? If something is moving upward, there must be an upward-pushing force, mustn't there? So what's the force? Antigravity?

Whoa! Watch your language. We don't use that word in this book. Science fiction is two shelves over to the left.

Surprisingly, there is no upward-pushing force. It's just that there is less downward-pulling force on the helium than there is on the air that surrounds it. The Earth attracts objects with a gravitational force proportional to the mass of the object. Because an average air molecule is almost four times as heavy as a helium atom, the air will tend to be pulled down past the helium, or—same thing—the helium will be observed to move upward past the air. If you were inside the helium balloon, you might be wondering, "Why is all that air rushing downward past me?" (See Einstein, relativity, and all that jazz.)

When you let go of your air-filled rubber ducky under the water in your bathtub, you're not surprised to see it zoom upwards, because it is less *dense* overall than the water. But helium and air, being gases instead of solids or liquids, are not as familiar to us; we can't see them, pour them, grab them, or throw them. But they are *matter* (substance) all the same, made up of atoms and molecules that are tugged upon by the Earth's gravitational field, and they respond in the same way that solids and liquids do.

The atmosphere itself is attracted downward, and the closer it gets to the Earth's surface, the more crowded its molecules become; the *denser* the air becomes. Conversely, the atmosphere becomes *less* dense as the altitude increases. So the simplest answer to your question about a released helium balloon would be this: If the balloon is made of a rigid material instead of elastic rubber, it would simply rise to the altitude at which the density of the atmosphere has decreased to match that of the balloon. And there, except for winds and turbulence, it would stay. (But see below.)

But when you let go of a helium-filled *rubber* balloon, several other things happen. As it ascends, it encounters changing conditions of both air pressure (density) and air temperature. At any given time, the balloon is a certain size because the outward-pushing pressure of the gas inside is counteracted by the inward-pushing pressure of the atmosphere outside (plus, of course, the inward-contracting tendency of the rubber). As the atmospheric pressure decreases with altitude, the outward-expanding tendency of the helium gas can prevail, and the balloon will expand if it can. So as the altitude increases, the balloon tends to get bigger. Hold that thought.

Now, what are the effects of decreasing temperature? We know that all gases will try to expand when heated and contract when cooled. That's because the molecules of a hot gas are bouncing around faster and pushing harder against any walls that are attempting to contain it. Our particular container of helium is ascending into colder and colder air: the average temperature of the earth's atmosphere decreases from about 70°F (21°C) at sea level to about –60°F (–51°C) at an altitude of 85 miles (137 kilometers). So as the balloon rises and gets colder, it will tend to shrink.

We now have two counteracting tendencies: an expansion due to the decrease in atmospheric pressure and a contraction due to the decrease in atmospheric temperature. Which tendency will win out?

The rules that govern the expansion and contraction of gases are well known; scientists lump them into a mathematical equation called the *ideal gas law*. Using this equation, one can actually calculate the effects of varying pressures and temperatures on a gas. If you do the calculations for our rising helium balloon (and I did), you'll find that the expansion due to the pressure decrease is a much bigger factor than the contraction due to the cooling.

So the net effect on the balloon is that it gets bigger and bigger as it rises until pop! The rubber can stretch no further and it bursts,

Nitpicker's Corner. Well, not exactly until doomsday. Other happenings can interfere with our rather neat picture. Our balloon may not even get high enough to explode, because the amount of helium in it isn't enough to carry its payload of rubber high enough and it'll settle out at some maximum altitude. Then, winds could blow it about for days, until enough helium has seeped out (helium atoms are extremely tiny as atoms go, and can diffuse right through the rubber) that the weight of the rubber brings it down. You've probably seen that happen to a balloon left on your ceiling for a couple of days.

eventually fluttering down into someone's picnic potato salad. The helium gas, now unfettered, just keeps scattering and rising through the atmosphere until some of it reaches a level at which the atmospheric pressure is as "thin" as it is, and that's where it will remain until doomsday.

Another monkey wrench in our simple mechanical analysis is that many helium balloons these days are made of aluminized Mylar, a tough plastic film coated with a very thin layer of aluminum, rather than rubber. A Mylar balloon will last a lot longer and go a lot higher without popping, because the Mylar can't expand like rubber. So it will rise until the atmospheric pressure decreases to the pressure of the helium inside the balloon, and it will rise no farther. Commercial jet airplanes have been known to spot them miles high, either quietly dozing or speeding along in the jet stream.

How many shiny Mylar balloons do you suppose have been mistaken for alien spacecraft?

Look! Up in the Sky! It's a Bird . . . It's a Plane . . . It's the Goodyear Prune!

Blimps, airships, dirigibles, lighter-than-air craft, whatever you call them, are filled with helium gas. But when they get heated and cooled by changing weather, the gas has to expand and contract. How do they allow for that? Does the whole airship expand and contract?

No, that would knock the sponsor's signs off the sides, and that would never do because today's blimps are nothing but flying billboards. Instead, they use a clever system of swapping helium and air back and forth.

The blimp is, as you've noted, essentially a big rubber balloon full of helium. The contraption floats in the air because the whole thing—helium, rubber bag, gondola, engine, crew, and joyriding local politicians—all together weighs less than an equal volume of air. It's exactly the same situation as a boat floating in water because its overall density is less than that of the water.

On a hot day with the sun beating down on it, there can be quite a pressure buildup inside the balloon. But they can't just vent some of that expensive helium out into the air. Moreover, what would they do when the bag cools and they need *more* helium to keep it from looking like a slowly sinking prune?

Here's the trick: There is a small, separate bag of air inside the big bag of helium: an air balloon inside a helium balloon. The balloons are arranged so that when the helium expands, it just pushes some cheap, old air out of the ship. And when the helium contracts, they make up for the shrinkage by pumping more air into the air bag. Or else they get one of those politicians to make a speech into it.

Why Astronauts Get Such a Warm Reception

Outdoors, the stronger the wind blows, the colder I feel. I think I understand that. But when a returning space shuttle plunges into the atmosphere, the passing air heats it up so much that they have to protect it from burning up like a meteorite, even though the air is a lot colder up there. How come, when the "wind" is strong enough, it turns from a cooling wind into a burning wind?

First of all, when it is very windy, the cooling effect on your skin has little to do with the evaporation of perspiration, in case that's what you were thinking. That effect peters out as soon as there is enough wind so that all the perspiration has already evaporated. A strong wind cools us because the stream of air molecules is carrying off heat from our bodies. The moment your skin heats up an adjacent air molecule, it is whisked away, carrying your hard-earned body heat along with it, and the faster the wind blows, the faster it carries off heat. Clothing protects you because it keeps those thieving air molecules from skimming alongside your skin.

As to the space shuttle: the first thing you have to do is to forget about "friction," the word that newspapers and magazines invariably use to "explain" the heat of atmospheric re-entry. Friction is a force that holds back the relative motion of two solids in contact; for liquids and gases, the word is utterly meaningless. The molecules of a gas are so far apart, with so much empty space between them, that they cannot exert a "holding-back" force. The only thing that gas molecules can do is to fly around and collide randomly with things, like a horde of houseflies dashing themselves madly against the display cases in a manure museum. (Sorry about that simile, but it describes exactly how gas molecules behave.)

Yes, the air is much colder and thinner at about 40 miles up, where the heating of a re-entry vehicle really begins to get serious. But when the "wind" is whooshing past the shuttle—or the shuttle is whooshing past the air (same thing)—at around 18,000 miles per hour, which is the speed at which the vehicle enters the atmosphere, we have quite a different situation from earthbound zephyrs. At 18,000 miles per hour, the shuttle is actually moving much faster than the individual air molecules as they flit randomly about.

The result is exactly the same as if the shuttle were standing still and the air molecules were bombarding it at their regular speed plus 18,000 miles per hour. That makes a total molecular speed that is equivalent to a temperature of several thousand degrees. (Think of a baseball bat—the shuttle—flying at 18,000 miles per hour toward a pitched ball: an air molecule. The effect is the same as if the ball had been hurled at its normal speed plus 18,000 mph.) The shuttle, then, feels as if it were being exposed to air at a temperature of several thousand degrees. If it hadn't been covered with a highly heat-resistant ceramic material that uses up energy by melting off, the shuttle would indeed burn up like a meteorite.

Even ceramics, however, cannot long endure at such high temperatures. Fortunately, the leading edges of the space shuttle are preceded by a shock wave, a layer of air molecules that pile up because they simply can't get out of the way fast enough. This layer of air acts as a front bumper on the vehicle. It soaks up the brunt of the heat energy by breaking down into a glowing cloud of atomic fragments and electrons: what scientists call a *plasma*. That's what makes the V-shaped "prow wave" that you see in those telephoto pictures on TV.

Drunken Rivers

When I look out the window of an airplane, I see rivers of many different shapes, ranging from mildly curved to the wildest snake-like twists. Some of them even have islands in the middle. How did all this come about?

Have you ever realized that rivers are the only bodies of water that travel? Just wondering.

You've heard that Nature abhors a vacuum. But it could also be said that Nature abhors a straight line. You've never seen a river flowing straight as an arrow for more than a short distance. If you were a river and had to search around to find the next spot that's a bit lower than where you are, you'd have to twist and turn too. All the square and rectangular plots of land you see from the airplane were laid out by humans, not Nature. And humans love straight lines, except on women's figures, where curves are keenly preferred.

We'll grant that Nature's waterfalls do drop in straight lines—down—as long as there are no obstructions or winds to divert them. But water flowing over land doesn't have that option. Its preference is indeed to flow generally downward, but it is subject to all sorts of surface obstructions and diversions along the way and must go over, around, and sometimes underneath them.

A river starts out in the mountains as a gully, a cleft in the surface that can fill with water. In rainstorms, the water in the gully runs downhill, eroding the soil and widening its path as it goes. Little gullies can become small streams in a few decades, rivers in a few centuries, and Grand Canyons in millions of years. As they run into one another and the water grows in volume and strength, it has to deal with all sorts of impediments in order to obey the rules of gravity. To deal with rocks in its path it makes rapids, and for unexpected sudden drops, it makes waterfalls. These features are usually formed in wooded areas, so you probably won't see them from the air.

In a river's middle age, the water runs faster, because falling objects gain speed (accelerate) as they are pulled down by gravity. Their increased power scours and erodes the land, picking up and carrying along anything its current is strong enough to move. In this way, rivers dig their own valleys. Most river valleys are V-shaped in cross section, at least when relatively young. That's because the waters at the sides of the river are slowed by dragging along the banks, so the water in the middle flows faster, has more power, and can pick up more rock and soil.

As the water gets down to more gently sloping land, it has room to stretch out and spread its power, thereby slowing down. Some of the stones and gravel it had picked up on the way down can't stay suspended anymore and drop out. That's where you may see sandbars or even islands made out of the dropped sand and gravel.

As the land gets flatter, the river spreads out even more and searches for better paths of least resistance. This often requires a right or left turn, so rivers bend. Just as on a racecourse, the water on the outside track has farther to go and must move faster to keep up with the rest. This faster water can pick up more sand and grit than the rest of the stream.

Now Sir Isaac Newton told us in his First Law that a body in motion will continue moving at the same speed in the same direction unless some outside force prevents it from doing so. When the fast-track, "outside" water hits the bend, it tries to keep going straight and crashes into the concave outer bank, wearing it away over the years by sandblasting it with its load of grit. Thus, the bank on the outer side of a river bend becomes steeper, even cliff-like, than the inner bank. You can see that best from low altitudes during takeoff or landing, or especially from the ground in a car or train. The older the river, the greater will be this difference in bank heights between the outer and inner parts of the curve.

The curvature of the outer bank bounces that fast water back across the stream to the opposite bank, where it may carve out a new bend in the opposite direction from the first, and the river begins to zigzag. Down in the flatlands, where the river is getting ready to empty into a bay or sea, it *meanders* in alternating left and right turns, tracing out a wavy path like a drunken river or a slithering snake. (The word meander comes from the ancient Greek name of a certain river in Turkey that, well, meandered a lot.)

Here and there, a meander may be so extreme that it makes a loop, sometimes so tight that it meets itself coming down and joins with it, closing the loop. The water trapped in the loop is shaped like part of the yoke of a team of oxen, and is called an *oxbow lake*. It may eventually dry up to form a flat island in the river. When you see an extremely loopy pattern, you're looking at an old river working its way toward its end.

Near the end of a river's path, where it empties into a bay or sea (for some reason, a river's outlet is called its mouth; *colon* would be more appropriate) it is no longer confined to a channel and spreads out into many small branches, like the branches of a tree. The slower water drops all of its sediments, making a broad, muddy plain. This geographic pattern is called a *delta* because of its triangular shape. (The uppercase Greek letter delta is a triangle.) Because of all the

---

You didn't ask, but . . .

Who made all those perfectly straight borderlines between farms that I see from airplanes over agricultural regions?

Obviously, they were laid out by surveyors, many of them a long time ago, at least by American historical standards.

In more ancient civilizations, natural features such as rivers, streams, and big rock formations served as landmarks and borders. But after the American Revolution, the United States became a new nation with an embarrassing wealth of land that had to be surveyed and distributed to its citizens for development.

The land was divided into squares of one square mile (640 acres) each. Subdivisions into smaller parcels gave rise to 320-acre half sections, 160-acre quarter sections, and 40-acre quarter-quarter (sixteenth) sections. To this day, farmers may refer to parts of their land as the "north forty," "back forty" and so on.

From an airplane, you will see these sections painted in different crop colors, depending on the time of year. In many cases, the boundary lines were originally marked by piles of stones that the farmer had to remove to enable plowing. Over time, the soil inside these lines of stones became homes to shrubs and trees. When you see a straight line of hedges or old trees, it may have been formed in this way.

Various plowing patterns decorate the sections: you can see strips, contours, and in flat parts of the country, spirals, which leave behind those perfectly circular areas you've undoubtedly wondered about. They might accurately be called crop circles, as distinguished from those patterns tramped into grassy fields by pranksters, trying to pass them off as mysterious prehistoric or alien art.

Contour plowing is done to encircle hills, by plowing small curves near the tops and graduating to broader curves as the slope diminishes. The resulting pattern looks just like a topographic map, which has contour lines drawn to connect points of equal elevation.

deposited soil and organic matter, river deltas are often quite fertile, and many ancient civilizations originated in those locations. When you fly over a river delta, you will see lots of greenery, even if the river is flowing out of a desert.

For aerial views of some impressive deltas, Google the deltas of the rivers Amazon (Brazil), Colorado (Mexico), Mississippi (Louisiana), Niger (Nigeria), and Nile (Egypt).

Or if it's not too far off your flight path from Detroit to Cleveland, ask your pilot to make a pass over the Ganges River delta, the biggest in the world.

Why Are the Oceans Salty?

All of our planet's oceans and seas are salty, and they seem to have roughly the same degree of saltiness everywhere. There must be a simple reason for that. Where did all that salt come from?

Many years ago, a sea captain bought a magic mill that turns out ground salt. But he wasn't told how to turn it off. He took it to his boat, which soon became filled with so much salt that it sank to the bottom of the sea, where the mill continues to grind out salt to this very day.

Oh, you don't buy that? (It's an old Norse fable.) Please read on. But you may skip the next paragraph if you know any chemistry at all.

What is a salt, anyway? When most people think of salt they think of sodium chloride, which is only one member of the large class of chemicals called salts. A salt is formed when an acid and a base neutralize each other. The result is usually a positively charged metal atom (a *cation*) bonded to a negatively charged nonmetal atom (an *anion*). For example, sodium chloride ($NaCl$) is the result of sodium hydroxide ($NaOH$) neutralizing hydrochloric acid (HCl). But seawater is far from a simple solution of sodium chloride in water. It contains many other salts and other kinds of chemicals.

End of chemistry lesson.

Where does all that stuff come from? Seawater has been described as a weak solution of almost everything. At least 70 of the 118 or so chemical elements have been found in it. You can't have a billion cubic kilometers of water splashing around the world without dissolving some of virtually anything with which it comes in contact. Seawater contains not only salt and dozens of other minerals, but lots of biological debris from the teeming life that lives in it. How the living things originally got there is a question to be pondered by evolutionary biologists, who, it must be said, don't really have any answers.

Here's the general consensus on how the minerals got there.

When the fiery young Earth cooled down and its surface solidified into a crust, rain and climatic temperature cycles began eroding the fire-born *igneous rocks* (from the Latin *ignis*, meaning fire) and wearing down the mountains. Rivers formed, dissolved some of the exposed minerals, and carried them down to the seas and oceans. But rivers don't have much ability to dissolve minerals, as is indicated by the fact that they are largely fresh, not saltwater. Nevertheless, over hundreds of millions of years of rain and other weather, rivers have been able to drag a lot of minerals down into the oceans. It is estimated that the rivers and streams flowing from the United States alone discharge 225 million tons of dissolved solids per year into its oceans and gulfs. Worldwide, rivers carry an estimated 4 billion tons of dissolved salts to the oceans every year.

Where do the rivers get their water? From rain, of course, but before that? From the oceans. The world's oceans have a surface area of 139 million square miles (361 million square kilometers). The sun beats down on all this water surface and on the land's surface of 57.5 million square miles (149 million square kilometers), causing the evaporation of a huge amount of water into the air. Eventually, this water falls back to Earth as rain, which flows down to the oceans in rivers, returning the water to its briny source. This sequence, called the *hydrologic cycle*, keeps recycling the Earth's water between the oceans and the atmosphere.

But minerals washed down by rivers are not enough to account for the saltiness—the *salinity*—of the oceans. The oceans' waters themselves can leach salts and other minerals from geological formations on and beneath their floors. Also, the recently discovered hydrothermal vents, which spew volcanically heated water from fissures in the sea floor, make their own contribution of minerals to the soup.

Note that all these processes have the effect of making the oceans progressively saltier as time goes by. Scientists believe that the oceans are indeed saltier now than they were millions of years ago.

Finally, why is sodium chloride by far the most prominent salt in the oceans? Every kilogram of seawater contains 10.8 grams of sodium and 19.4 grams of chloride—about the same amounts throughout the world. That's eight times the amount of magnesium and seven times the amount of sulfate, the next most abundant cation and anion. The reason is that salts have various degrees of solubility—the maximum amounts that can dissolve in water—and sodium chloride is 1.4 times as soluble as magnesium sulfate. So when a river sucks the minerals out of its course, it's going to suck up and dump in the ocean more sodium chloride than any other salt. And that's just the way the fish like it.

CHAPTER 6

Water, Water Everywhere

Water is the most abundant chemical compound on Earth. It covers about 75 percent of our planet's surface, making the Earth look blue and white from space. (The white clouds, of course, are also water.) The total amount of water on Earth, including oceans, lakes, rivers, clouds, polar ice, and chicken soup, amounts to 1.5 billion billion tons. In fact, we ourselves are more than half water: a typical 150-pound male is about 60 percent water; females average closer to 50 percent; fatter people are lower. Babies can be as high as 85 percent water, not counting the diapers.

Water possesses some of the most unusual properties of any chemical in the universe, and yet it is so familiar to us that we take it completely for granted. But what is really happening when we boil it, freeze it, float on it, or perspire it? In this section, we'll peek beneath the surface of our everyday encounters with this most remarkable liquid.

How Sweat It Is!

I know that people sweat as a mechanism for keeping cool, because evaporating perspiration has a cooling effect. But why? Just because a liquid is evaporating, why should its temperature go down?

We notice that our sweat glands are exuding a liquid—water containing a little salt and urea—onto our skin only at certain times, such as (*a*) when we're hot, (*b*) when we're exerting ourselves strenuously, or (*c*) when we're about to deliver an important speech and can't find our notes. In reality, however, our perspiration process is always working, even in cold weather; it's an essential mechanism for keeping our body temperature constant. In situations like *a, b,* and *c* above, the perspiration is being generated faster than it can evaporate, so we notice the buildup of actual moisture on our skin.

Dogs, being called upon much less frequently to deliver speeches, do not have sweat glands on their skin (except, oddly, on the pads of their feet). So they hang out their extraordinarily long tongues and pant, which hastens the evaporation of saliva, thereby cooling the air supply to their lungs. Other animals sweat to varying degrees. Pigs do indeed "sweat like pigs" on occasion, although they also like to cool off by wallowing in mud, as do elephants and hippopotamuses. Not much different, actually, from our own habit of taking a quick dip in the pool.

But what, exactly, is evaporation? It is the process in which certain highly energetic molecules at the surface of a liquid use their energy to depart from their brethren and fly off into the wild blue yonder. As more and more molecules depart, the amount of remaining liquid, of course, diminishes. And because the departed molecules are the highest-energy ones, the average energy—or temperature—of the remaining liquid goes down. So we wind up with both drying and cooling. You've seen it happen dozens of times: wet floors dry up and laundry dries on the clothesline.

If we want to hasten evaporation, we can do two things: heat and blow. Heating the liquid gives more of its molecules the energy they need to escape. Hence, hair driers and hand driers in public rest rooms heat the air, which in turn heats the wet surfaces and speeds the evaporation.

Blowing air also speeds evaporation. Think of a water molecule attempting to evaporate from the wet surface. Unless the air adjacent to the surface is bone dry, there will always be some water vapor in the air; that's what *humidity* is. And just as some water molecules are leaving the liquid to go off into the air as vapor, some vapor molecules in the air are likely to be leaving the air and returning to the liquid. It's a two-way street. The balance between liquid-to-vapor traffic and vapor-to-liquid traffic determines what the net result will be: if there are gazillions of water-vapor molecules in the layer of air adjacent to the liquid's surface—that is, if the air's humidity is high—then vapor-to-liquid traffic will prevail by sheer numbers, and evaporation will be inhibited. But if a wind or breeze is sweeping away many of these vapor molecules, then liquid-to-vapor traffic can prevail and evaporation is enhanced.

Blowing on hot soup to cool it off is a classic, though inelegant, application of this principle. And you'll feel cold coming out of the bath if the room is drafty, even though the air temperature may be quite comfortable. Outdoors, you'll always feel colder when it's windy. The "wind chill factor" that northern weather broadcasters love to frighten us with during the winter is an attempt to take this phenomenon into account. Unfortunately, it only applies when you're naked.

Okay, so why should the exodus of water molecules lower the temperature of the remaining liquid, and therefore the temperature of whatever it's in contact with? This may sound almost spooky, but the process of evaporation is highly selective; it preferentially picks out and removes the faster (hotter) molecules, leaving the cooler (slower) ones behind. Here's how.

The molecules of any liquid are in constant motion: sliding around, jiggling back and forth, darting about, colliding with one another, and generally acting like a soup bowl full of live ants. The higher the temperature, the faster the molecular motion (and the faster ants move, if you really want to know). In fact, that's what temperature *is*: a measure of the average *kinetic energy* (energy of motion) of all the molecules in the substance.

The important word here is *average*, because at any given temperature, the molecules are by no means all moving at the same speed. Some may be moving very fast because they've just been kicked by a collision with another molecule. Meanwhile, the molecule that kicked it is moving slower because it has just given some of its energy to the molecule it hit. Go to the nearest pool table and you'll see that the cue ball slows down substantially when it hits another ball, while the struck ball goes sailing away at high speed. But the average energy of the two balls—their *temperature*—remains the same.

Now at the surface of a liquid, which molecules do you suppose are most likely to leap into the air and evaporate? The highest-energy ones, of course. And that will lower the average energy—the temperature—of the molecules that are left behind. Thus, as a liquid evaporates, it cools down.

But that's not the end of the story. The cooling can't go on without limit. Did you ever see an evaporating puddle spontaneously freeze itself solid? No, what happens is that as soon as the evaporating liquid begins to cool a bit, heat flows in from its surroundings and replenishes the population of high-energy molecules, which in effect keeps the temperature constant.

"Aha!" you say. "Then we're back to square one. If the evaporating liquid never gets a chance to stay cold, why does evaporating sweat cool me?"

Well, where do you think that replenishment heat has to come from? No place but your skin. As the evaporation proceeds, then, the sweat layer itself never does get a chance to cool down very much, because it keeps taking up heat from your skin and throwing it off into the air in the form of its hottest molecules. The sweat is just a go-between, helping your skin to throw heat away.

TRY IT   *Put some isopropyl, or rubbing alcohol, on your skin and you'll feel a much greater cooling effect than you get with water, especially if you blow on it. That's because the "hotter" alcohol molecules are leaving at such a rapid rate that they out-pace your body's ability to warm your skin back up to body temperature.*

Ethyl chloride is an extremely volatile liquid whose molecules don't really want to have much to do with each other and who are just dying to leave home. It evaporates about a hundred times faster than water. Put some ethyl chloride on your skin and it'll get so cold that it numbs your sensations. Doctors use it as a local anesthetic for minor skin surgery.

The rate at which liquids evaporate depends on how tightly they are bound together in the liquid. In a liquid where the molecules aren't very strongly attached to each other, they can leave the crowd more easily and the liquid evaporates more rapidly. Some liquids evaporate so fast—are so *volatile*—that replenishment of heat from the surroundings can't keep up. In that case, the temperature of the liquid really does go down.

Archimedes at Sea

How can a 100,000-ton aircraft carrier possibly float on water? I know that if it were a solid chunk of steel it would sink, but it isn't solid; it's hollow. But how does the water underneath know that?

The pat answer to the everyday puzzle of why things float invariably goes like this: "According to Archimedes' principle, a body immersed in a fluid is buoyed up by a force equal to the weight of the fluid displaced." Perfectly correct, of course, but just about as illuminating as a firefly wearing an overcoat.

Obviously, the water underneath a ship has no information as to whether the object pressing upon its surface is a solid lump or is a sea-going Swiss cheese (except for holes in the hull, which we'll get to). Nevertheless, most of our experience with floating things, from canoes to plastic foam, makes us believe that hollowness, air spaces in the interior of an object, is somehow necessary. It is not. Hollowing things out is just one way of making them lighter. Light things float

---

You didn't ask, but . . .

How about submarines? They prefer to float sometimes and to sink sometimes. How do they change their buoyancy?

Very simply. They change their amount of internal air space, thereby changing their density. You want to dive? You let water into your ballast tanks to replace some of the air. You want to surface? You blow the water out with compressed air. It gets a bit tricky in reality, though, because the density of the seawater actually varies a bit, depending on depth, temperature, and salinity (saltiness). The density of the submarine therefore has to be continually adjusted.

Nor did you ask this, but . . .
Exactly why does a hole in the hull of a ship make it sink?

Water rushes in through the hole because it is under a pressure, depending on how deep below the surface the hole is: the lower the hole, the more forcefully the water rushes in. As the water enters the ship, it replaces an equal volume of air, thereby increasing the ship's weight and density. When enough water has entered to make enough extra weight to overcome the buoyancy force, down she goes.

---

and heavy things sink. Which is just what you would have expected if that old Greek hadn't muddied the waters, so to speak.

The question is, just how light does an object have to be in order to float on water? And the answer is, lighter than an equal volume of water. If an entire ship, considered as a huge, complex conglomeration of metal, wood, plastic, air, and rats, weighs less than an equal volume of water. That is, if the ship's density is less than the density of water, it will float. A solid block of wood floats because its density is only about 60 percent of the density of water. No hollowing required.

But if we want to float 100,000 tons of mostly-steel aircraft carrier, we'd better do some serious hollowing to get its overall density down. That's no problem, of course, because the hollows are convenient spaces to stow such necessities as airplanes and sailors. If the whole ship were melted down into a solid block of steel and stuff, it would sink like a rock.

To find out why a floating object has to be less dense than water, let's do a little experiment. Let's lower the 100,000-ton nuclear-powered aircraft carrier USS *Theodore Roosevelt* (the world's largest) very gently into a rather large bathtub of water. Gravity does the lowering job for us by pulling the ship downward into the water with a force equal to its weight. (That's what weight *is*.) But as it enters the water, the water has to make way for it. The ship must push some water aside and upwards against water's natural gravitational preference for maintaining a low profile. So as gravity pulls the ship down, the ship forces water up against gravity. Notice the level rising in the bathtub?

How much water can eventually be forced upward against gravity? Only as much upward force—known as *buoyancy*—as the downward force of gravity on the ship. In other words, the weight of the water

You didn't ask this either, but . . .
According to Archimedes, there is a buoyant force that pushes upward against any object that is placed in water. Where does that force come from?

If you doubt that the water exerts an upward force, try to submerge an inflated balloon in the bathtub. You'll feel a substantial upward push that resists your downward push.

When we lowered Teddy Roosevelt into the bathtub, the water level rose; it got deeper. As every diver knows, deeper water means higher pressure. This increased pressure is present everywhere throughout the water in the bathtub, because water is not compressible; it cannot cushion or absorb a force, the way a spring or a piece of rubber can. The water must transmit its increased pressure in all directions to everything it is in contact with, including the ship's hull. All the north-east-south-west horizontal pushes on the hull cancel each other out, leaving only an uncancelled upward push on the bottom. This is the pressure that pushes the ship upward against gravity. Voilà! Buoyancy.

Okay, I know what you're thinking. Aircraft carriers operate much more frequently in oceans than in bathtubs. Am I telling you that the level of the ocean was raised when Teddy was launched? I certainly am. Spread that 100,000 tons of water over the surface of the entire Atlantic Ocean, though, and it comes to a pretty meager rise, quite unlikely to flood any beachfront property in Florida. Nevertheless, it's a volume of water equal to the submerged volume of the ship's hull up to the water line, and a buoyant force equal to that weight of water is still operating on the ship.

Archimedes had only a rubber ducky in his bathtub, not an aircraft carrier, so as the story goes, he used his own body. He filled his bathtub to the brim, climbed in, and realized that the weight of the overflow water on the floor must be the same as his loss of weight, his buoyancy in the water.

History does not record the reaction of his landlady.

that is lifted or displaced will be equal to the weight of the ship, exactly as Archie said. As the ship settles into the water, the upward push of the water eventually becomes equal to the downward pull of gravity on the ship, and it stops settling down. By God, it's floating!

How far up the sides of the ship will the water go? Each cubic foot of displaced water must have been displaced by exactly one cubic foot of ship that's immersed in the water. That is, the volume of ship below the water line is the same as the volume of 100,000 tons of seawater. Because seawater is more dense than the ship (or else the ship wouldn't be floating), 100,000 tons of seawater occupy less volume than 100,000 tons of ship: less than the ship's entire volume. Luckily, then, the water needn't rise so far as to envelop the entire ship.

—All because the ship's overall density is less than that of seawater.

TRY IT

Seawater is about 3 percent denser than fresh water. A ship in ocean water is therefore buoyed up by a 3-percent greater force than a ship in a lake, and it therefore floats a bit higher.

The Dead Sea and the Great Salt Lake are so dense from their high salt contents that their buoyancy is astonishing. Try floating in one of them if you ever get the chance. You'll only sink in a few inches. It's

an amazing sensation, worth going out of your way a few thousand miles for.

Fish Gotta Swim

While snorkeling around in the water, I saw a shell on the bottom that I wanted to collect. I tried to dive down, but even with flippers, it was devilishly hard to force my body down that deep. It infuriated me that dumb little fishes were diving all around me. What have they got that I haven't got?

The trouble is not what they've got, but what you've got: lungs.

In order to feel at home when suspended in seawater in precisely neutral buoyancy without sinking or rising, a fish or any other object must have exactly the same overall density as the seawater. If its density is higher, it will sink to the bottom. If its density is lower, as most humans' is, it will float. Ships, of course, are scrupulously designed to achieve the latter condition.

Bone, with its density of 1.750 g/cm^3 (grams per cubic centimeter) and muscle at 1.060 g/cm^3 are both denser than seawater at its average surface density of 1.025 g/cm^3. (The density of pure water is 1.000 g/cm^3.) So most animals will be sinkers unless they contain a lot of lighter stuff, such as fat at 0.909 g/cm^3 or air at 0.0013 g/cm^3.

We humans contain a lot of air in our lungs; the lung capacity of an average adult male is about 6,000 cm^3. That's what makes most people float, even in a freshwater swimming pool. Our lungs lower our overall density so much that our bodies are more buoyant than many kinds of wood.

On the other hand, a short, lean, muscular friend of mine was a natural sinker. We used to throw him into the pool and watch him glide, motionless and grinning, all the way down to the bottom.

Fish do their oxygen–carbon dioxide exchange through gills, rather than lungs, so lung-buoyancy isn't a problem. But they generally have swim bladders, little gas-filled bags. Fishes' swim bladders make up only about 5 percent of their entire volume, though, while our lungs fill most of our chest cavities. Even if a fish happens to be temporarily denser than seawater, it can avoid sinking by swishing its tail or flipping its fins.

But alas! You're just not as proficient in the art of fin-propulsion as the fishes are. Even if you were, you would still have to work much harder, because you are burdened by those huge internal water wings called lungs.

You didn't ask, but . . .

If a fish's density is just right for staying suspended in the water, how does it manage to rise or sink whenever it wants to change its depth?

Of course, it can always use its tail and fins to get to wherever it wants to go, but that's just a temporary solution. What it would really like to do is adapt to the pressure of the new depth, so that it can maintain its neutral buoyancy and rest there without constantly having to struggle up or down. It does that by adjusting its swim bladder.

When a fish travels to deeper water, the pressure on him or her increases because there is more water pressing down from above. This higher pressure compresses the swim bladder, making the fish denser than the precise value it needs for neutral suspension. In order to remain effortlessly at that depth, it would have to re-expand its swim bladder. Conversely, when a fish ascends to shallower depths, it would have to compress its swim bladder in order to stay in neutral suspension without having to swim.

This was the simple mechanism we used to believe. But it was found that fish just don't have the necessary expansion and contraction muscles to do that. Surprisingly, what they do instead is change the amount of oxygen gas in the bladder. By adding or removing gas from the bladder, a fish can readjust its density to the exact density of the water and remain effortlessly at that depth without having to swim too much, no matter what the water pressure has done to its bladder size.

Where does a fish get the extra gas that it needs when it wants to live at a higher "altitude"? It takes oxygen out of its bloodstream and secretes it into the bladder. Where does it stash the extra gas when it wants to live at a higher "altitude"? It absorbs some of the oxygen from the bladder back into the bloodstream. Ingenious!

Some poor fish don't have swim bladders. They are slightly denser than seawater and have to keep swimming to stay off the bottom. Mackerels and some species of tuna begin to sink as soon as they slow down. But the turbot just gives up and stays on the bottom.

So if you have to work hard in order to dive, take comfort in the fact that some fish have to work hard to keep from sinking.

How to Bend a Fish

A usually reliable source told me that fish can get the bends, just as scuba divers do when they stay down too long. Okay, I feel silly for asking, but how long can a fish stay under water without getting sick?

Fortunately, it isn't necessary to answer that question, because divers and fish don't get the bends (more accurately known as *decompression sickness*) from staying down too long. They get it by coming up too fast.

When the water pressure on a diver's body is reduced too fast, gas can bubble out of his or her bloodstream. That hurts, to say the least. And yes, the same thing can happen to a fish, but usually not from swimming upward too fast. It happens because of changes in the characteristics of the water itself.

Air dissolves to some extent in water and in watery liquids such as blood and body tissue fluids. That's just great for the fish, of course, because they live on the oxygen that is dissolved in the water. But the nitrogen, which is the major (78 percent) component of air and which is inert and useless to physiological processes, also dissolves in water and in blood. Ordinarily, this causes no problem for either fish or man, because we extract the oxygen that we need for metabolism and throw away the rest via gill or lung. But when there is too much dissolved air in our bloodstreams, we may not be able to "undissolve" the excess nitrogen fast enough, and it can collect into actual bubbles of gas, blocking the circulation and destroying local tissues.

The amount of air that will dissolve in water at a certain temperature depends on the pressure: the higher the pressure, the more gas will dissolve. When a human diver goes down, the increased water pressure forces more oxygen and nitrogen into his or her bloodstream via the lungs. The oxygen is no problem; the blood's hemoglobin eagerly latches onto whatever oxygen it can find and delivers it to the cells; that's its job. When the diver surfaces and the pressure decreases, it would be very nice if the excess dissolved nitrogen could depart via the way it came in: through the lungs. Unfortunately, that can be a very slow process. Instead, when the pressure is reduced too quickly, the excess nitrogen gas can bubble out of the blood, just as the carbon dioxide does when you release the pressure by opening a bottle of soda pop.

The answer for divers is to come up slowly. Give the nitrogen a chance to leave the bloodstream gradually, molecule by molecule, and to exit via the lungs.

If a fish were to swim rapidly upward from a great depth to the surface, the same thing might happen, except for two differences: one, fish have more sense than to do that and two, something even more drastic would happen: the fish's swim bladder would expand so much that it would crush the fish internally and kill it.

But we said that fish can get the nitrogen-bubble bends, and they can. Here's how.

Suppose that a fish is happily acclimated to its environment, swimming around in water at a certain temperature that contains a certain amount of dissolved air, depending on the depth and pressure. Its bloodstream will have adjusted itself to that same amount of nitrogen. Now suppose that the fish wanders into water that for some reason (we'll get to the reasons) contains a lot more dissolved nitrogen than is normal at that temperature and pressure. Before long, its blood will also acquire that same abnormal amount of dissolved nitrogen. This is a precarious condition to be in, though, because at any moment, some of that excess nitrogen can pop out as a bubble and the fish will get the bends. It can only relieve itself by heading for deeper depths, where the added pressure will push the bubble back into the blood.

How does a fish find itself in water that contains an abnormal amount of nitrogen? It turns out that it needn't have anything to do with depth or pressure.

For one example, a fish may be swimming in a river that contains the standard atmospheric pressure's worth of dissolved nitrogen, when it happens upon a region of warmer water that has just been discharged by a factory or power plant. (Power plants inevitably throw away a lot of waste heat.) By rights, the warmer water should contain less nitrogen, not more, because gases dissolve to a lesser extent in warm water than they do in cold. But if the plant's water had never been given enough time to lose some of its nitrogen when it was heated—and remember that the evolution of nitrogen can be a slow process—then it will still be carrying more nitrogen than is normal for the river's atmospheric pressure. The poor fish finds itself swimming in abnormally high-nitrogen water, and it gets the bends. That's one way in which power plants kill the fish in a river "merely" by discharging warm water into it.

Another example: have you ever bought a couple of goldfish, taken them home, put them in a bowl of nice, fresh water and then watched them get sick and die? Well, here's what might have happened. Your tap water has lots of dissolved air in it because it is cold and it was probably even sprayed into the air at the waterworks to aerate it. Then you put it into the fishbowl, where it slowly warms up to room temperature. But it may still retain its cold-water load of nitrogen because, as stated above, evolving excess nitrogen can be a very slow process. The water then will still contain an abnormal load of nitrogen when you put the fish in it. Bends and death ensue.

Can anything be done about the power plant fish-kills and the thousands of goldfish murders that are committed every day? Yes, and it's fairly simple. Just let the water stand for a long time before dumping it into the river or the fishbowl. Standing will permit excess nitrogen to escape, and the water will come down to the just-right amount of nitrogen content for its temperature and pressure, and therefore the just-right amount for a fish at that temperature and pressure.

You didn't ask, but . . .
How do fish in deep ocean water get their oxygen? How much oxygen can there possibly be down there, with the atmosphere so far above?

The oxygen doesn't come only from dissolved atmospheric air. You're forgetting about plants, which breathe in carbon dioxide and breathe out oxygen. Oceans contain an abundant variety of plant life, and the oxygen emitted by the plants dissolves directly into the water. By constantly swimming and passing large amounts of water over its gills, a fish can "vacuum up" a lot of oxygen, even if it isn't present in very concentrated amounts.

In areas where there don't happen to be enough plants to supply the fishes' breathing needs, they just take their business elsewhere. Fish, like teenagers, tend to congregate where they don't have to work too hard.

Blowing Bubbles in the Air

Why are soap bubbles round?

Let's put it this way: you'd be pretty surprised if they were square, wouldn't you? That's because all of our experience since we were babies tells us that Mother Nature prefers smoothness. There just aren't many natural objects that have sharp points or jangling angles. The major exception is mineral crystals, which occur naturally in beautifully angled geometric shapes. That may be why the world's woo-woos believe that crystals and pyramids are endowed with supernatural powers.

But that is metaphysics, not science. Bubbles are spherical because there is an attractive force called surface tension that pulls molecules of water into the tightest possible groupings. And the tightest possible grouping that any collection of particles can achieve is to pack together into a spherical ball. Even the stars and planets know that. Of all possible shapes—cubes, pyramids, irregular chunks—a sphere has the smallest amount of outside surface area.

As soon as you release a bubble from your bubble pipe or from one of those more modern gadgets, *surface tension* makes the thin film of soapy water assume the smallest surface area that it can; it becomes a sphere. If you hadn't deliberately trapped some air within it, the soapy water would continue to shrink down to a spherical droplet, as rain drops do. Surface tension, as a contracting force, arises because the molecules on the surface of a liquid are attracted to all their brethren *below* the surface, but not *above* the surface—there are no attractive molecules there—so there is a net, unbalanced force tending to pull them inward. That makes a sort of skin on the bubble, like the rubber skin on a balloon, tending to pull it inward.

But the air on the inside is pushing *outward* against the water film; all gases exert a pressure on their captors because they consist of freely flying molecules that are banging up against anything in their way. In a bubble, the inward, contracting forces of the surface tension are exactly balanced by the expanding pressure of the air inside. If they were any different, the bubble would either shrink or expand until they were equal.

Try to blow more air in to make a bigger bubble? That makes a higher air pressure inside. All the water film can do to counterbalance the increased outward pressure is to expand its surface, making more inward-directed surface-tension forces, like a rubber balloon's effort to contract.

You didn't ask, but . . .
Why do we need soapy water to blow bubbles? Why can't we use plain water?

In the strength of its inward-directed surface-tension force, water is the champion of all liquids. Its surface tension is so strong that water resists being stretched outward at all, even into the three-dimensional shape of smallest surface area: a sphere. Water knows that it can have an even smaller amount of surface area by simply lying flat like a puddle and refusing to extend up into the third dimension at all. So pure water won't make bubbles of any shape—at least, not bubbles that will last for more than an instant. Soap has the effect of reducing the surface-tension skin of water. It weakens it enough so that the water's "skin" can be stretched upwards into three dimensions, making a stable, long-lasting, three-dimensional bubble.

Alcohol, on the other hand, has such a low surface tension that it won't make bubbles worth a damn. It would be like trying to blow bubbles with ordinary chewing gum that has virtually no elasticity.

So the soap bubble very cooperatively grows in size. But it must get thinner in the process, because there is only so much liquid to go around. If you're being piggy about it and keep blowing more air in, the film eventually won't have enough reserve liquid to spread out into a bigger surface, and the ultimate catastrophe occurs: your bubble bursts.

Exactly the same thing happens with bubble gum, except that instead of surface tension as the inward, contracting force, it's the elasticity of the rubber in the gum. (Yes, rubber.) Elasticity, like surface tension, means, "Let's always try to assume the smallest possible shape."

Building a Better Wetter

Are all liquids wet?

No, all liquids are not wet. Even water is not always wet. It depends on who or what is the "wet-ee."

Make this inquiry of a linguist, however, and you'll be told that it is a foolish question. The word *wet* is so intimately related to the word *water* in the roots of our language that "wet" has always meant

Nitpicker's Corner. Actually, wetness is a relative term. Some liquids are wetter than others; they will spread out and flow more readily over the surface that they are wetting.

Surprisingly, water isn't a very good wetter, as liquids go; ethyl alcohol, for example is much wetter than water. That's because water's molecules adhere to each other so strongly (by *dipole-dipole attraction* and *hydrogen bonding*) that they tend to ignore other nearby molecules and won't adhere to them very readily even if they do have the right kind of molecular attractiveness.

"coated with water." Water is therefore "wet" by definition, and the opposite of "wet" is dry, which means "without water:" no longer wet.

But language is a fickle facsimile of fact. The reason for the linguistic intimacy between water and wetness is simply that no other liquids were known by our primitive ancestors when they needed a word to describe the way you look when you come out of a river. After all, water is not only the most abundant liquid on Earth, it is the most abundant chemical compound of any kind. Even today, most people would be severely challenged to name two or three other liquids. Things like blood and milk don't count, of course, because their liquid parts are still water.

Innumerable other liquids do exist, however. In principle, any solid material can be melted into a liquid by heating it, and any gas can be condensed into a liquid by cooling it. It just happens that water exists in its liquid form over most of the temperature range at which life also exists. That's no coincidence, of course, because life began in the water, and liquid water is still essential to all forms of life.

Why, though, is this ubiquitous liquid *wet*? Why does it stick to us when we emerge from the river? Our primitive ancestors would have loved this explanation: *it sticks to us because it likes us.*

TRY IT

Dip a candle into a glass of water and you will see that water isn't necessarily wet. Water is sometimes "wet" and sometimes not, depending on what material we are tempting it to stick to.

Putting it a little more scientifically, water molecules will adhere to those substances whose molecules hold some form of attraction to them. If there were no attraction between the molecules in a drop of water and the molecules at the surface of our skins, the water would just roll off. Our job, then, is to find out what those attractive forces may be.

At several other places in this book, we talk about the fact that water molecules are *polar*, and attract each other like tiny magnets. Water molecules are attracted to each other also by *hydrogen bonding*. If an

TRY IT

If you ever get a chance to dip your finger into a pool of metallic mercury, which is liquid at room temperature, you will note that it comes out as dry as that candle that you dipped in water. But dip a piece of clean copper or brass into the mercury and it will wet it eagerly, because metal atoms all have similar attractive forces and tend to stick together. If you have ever done any soldering, you know that the melted (metal) solder wets the metal parts that you are trying to join together.

Bar Bet. All liquids aren't wet, and even water is sometimes dry.

alien substance comes along whose molecules are also polar or are also subject to hydrogen bonding, the water molecules will be attracted to them as if to their own. In other words, water will wet that substance.

Most proteins and carbohydrates, including the proteins in our skin and the cellulose in wood, paper, cotton, and other vegetable matter, are made of molecules with the right characteristics to make water molecules want to snuggle up to them; they will therefore be wetted by water. Furthermore, animal and plant cells already contain water, so they will be attractive to other water molecules and would be wetted by them anyway. Other kinds of substances, however, such as oily or waxy materials, don't have either of the two necessary molecular characteristics to be wet by water.

What about other liquids? Are they always "wet"? We might wonder about such liquids as grain (*ethyl*) alcohol or isopropyl rubbing alcohol, gasoline, benzene, olive oil, and even liquid metals such as mercury. Like water, these liquids will wet materials to whose molecules they are attracted by a mutual attractive force. As far as human skin is concerned, the first four can find enough in common with "skin molecules" to adhere to them, and these liquids will wet you. But the atoms of metals have nothing in common with your "skin molecules" and won't wet them at all.

TRY IT

Sprinkle a few drops of water onto your umbrella and they will roll off, unless you force them to wet the material by rubbing them in with your finger. Sprinkle some alcohol onto the umbrella though, and it will soak right in. Umbrella fabric has been coated with a waxy or oily material that water won't wet, but alcohol will. On the other hand, there are certain substances that, when added to water, will make it a better wetter. Soap is the most common one.

Does Hot Water Freeze Faster than Cold Water?

Can hot water really freeze faster than cold water? Some people swear that it does, while others appear ready to clobber them for saying so. Is there a definitive answer?

Well, yes and no.

This controversy has been raging ever since the early seventeenth century, when the eminent philosopher, statesman and scoundrel Sir Francis Bacon became a charter member of the Betcha-The-Hot-Water-Freezes-First Society.

The only conscientious answer to this puzzle is, "It depends." It depends on precisely how the freezing is being carried out. Freezing water may sound like the simplest of happenings, but there are many factors that can affect the result. How hot do you call hot? How cold do you call cold? How much water are we talking about? What kinds of containers are they in? How much surface area do they have? How are they being cooled? Exactly what do we mean by "freezes first"—a skin of ice on the surface or a solid block? And how much money or prestige is riding on the outcome? This last consideration has been known to affect the judgment of the most careful experimenter, consciously or not.

Let's listen to some of the yea-sayers and naysayers.

NAY: It's impossible! Water has to be cooled down to 32°F (0°C) before it can freeze. Hot water simply has further to go, so it can't possibly win the race.

YEA: Yes, but the rate at which heat is conducted away from an object is greater when the temperature difference between the object and the surroundings is greater. The hotter an object is, then, the faster it will cool off, in degrees per minute. Therefore, heat will be leaving the hot water faster and it will be cooling faster.

NAY: Maybe. But who says the heat is leaving by being conducted away? There's also convection and radiation, you know. And anyway, that would mean only that the hot water might catch up with the cold water in the race toward 32°F (0°C). But it could never pass it. Even if the hot water gets to be the same temperature as the cold water, they will thereafter continue to cool at the same rate. At best, they'll freeze at the same time.

YEA: Oh, yeah?

NAY: Yeah!

Having reached the point of diminishing returns of rationality, we may mediate the discussion by stating that thus far, the nays have it. Clearly, under absolutely identical, controlled conditions, hot water could never freeze faster than cold water. The problem is that hot water and cold water are inherently not operating under identical conditions. Even if we had two identical, open containers being cooled in exactly the same way, there are several factors that could possibly bring about a hot victory. Here are some of them:

• Hot water evaporates faster than cold water. If we start with exactly equal amounts of water (which is essential, of course), there will be less water remaining in the hot-water container when it gets down to rug-cuttin' time at 32°F (0°C). Less water, naturally, will freeze in less time.

If you think that evaporation can have only a trivial effect, consider this: At the typical temperatures of hot and cold household tap water (about 140°F (60°C) and 75°F (24°C), respectively), the hot water is evaporating almost seven times as fast as the cold. Over a period of an hour or two, a container of hot water can be substantially diminished by this rapid evaporation. Of course, as the hot water cools down, its rate of evaporation will gradually decrease; nevertheless, on the way down it may well have lost a substantial amount of water.

• Water is a very unusual liquid in many ways. One of these ways is that it takes quite a lot of heat, relatively speaking, to raise its temperature by each degree. (Water has a high *heat capacity*.) Conversely, quite a lot of cooling is required to lower its temperature by each degree. If there is only slightly less water in a container, then it may require substantially less cooling to get it down to freezing temperature. If the originally hot container has lost even a little bit of its water by evaporation, it may reach the freezing temperature quite a bit sooner than the water in the other container. That is, it could actually overtake the cold water and get to the finish line first. Moreover, once it is at the freezing point, it's not yet frozen; it must have a great deal of extra heat removed from it to make it actually freeze into ice: 80 calories for every kilogram. So again, a little less water can mean that a lot less cooling is necessary to freeze it.

• Evaporation is a cooling process. The faster-evaporating hot water will therefore be adding some extra evaporative cooling to whatever cooling process is operating on both containers. Faster cooling can mean faster freezing.

• Hot water contains less dissolved air than cold water does. Anything, including a gas, that is dissolved in water makes the water freeze at a lower temperature. The more air (or anything else) that is dissolved in water, the lower is the temperature that it must be cooled to, in order to freeze. Having less air in it, the hot water doesn't have to be cooled to as low a temperature as the cold water does, and can freeze sooner. However, this often-quoted argument doesn't hold water, so to speak. The lowering of the freezing point due to dissolved air amounts to only a couple of thousandths of a degree. Nevertheless (there's always a nevertheless), many people claim that when the water pipes freeze in an unheated house in the winter, it is usually the hot water pipes—containing previously heated water—that freeze first.

All things considered, then, it is quite possible that under some circumstances a bucket of hot water, left outside in the winter, will freeze faster than a bucket of cold. The claims of Canadians that they have seen it happen many times can be believed, even by scientists who "know better" and by other skeptics. The biggest and most probable effect in this case is the loss of water by evaporation. Extensive research, however, has not yet revealed why Canadians leave buckets of water outside in the winter.

But there are still a couple of hefty monkey wrenches in the works. First of all, a container of water doesn't cool uniformly throughout, until it suddenly freezes. It cools irregularly, depending on the shape and thickness of the container, what it is made of, the prevailing air currents, and several other variable factors. The first skin of ice to form at the surface of the water, therefore, may be a bit of a fluke, and may not be signaling that the rest of the water is ready to freeze. (The first ice to form will invariably be at the surface of the water.)

Second of all, believe it or not, water can be chilled well below its freezing point without freezing. It will be super cooled, yet it will not crystallize into ice unless some outside influence stimulates it to do so. The molecules may be all ready to snap into their rigid ice-crystal formation, but they need some final encouragement, perhaps in the form of a speck of dust that they can gather around, or maybe an irregularity on the wall of their container.

In view of these uncertainties, precisely when can we say that a given container of water has "frozen"? Our two buckets of water are racing without a clearly defined finish line labeled "frozen."

All things considered, the best we can say is, "Yes Virginia, hot water can freeze faster than cold water. Sometimes."

If you are tempted to run right into the kitchen to fill two ice-cube trays, one with hot water and one with cold, and put them into your freezer to see which one freezes first, don't bother. There are just too many uncontrolled variables. You can get one result one time and the other result the next. That's the problem with people who say, "I know it works, because I tried it" in reference to anything from freezing water to curing warts. You just have to examine everything that could possibly affect the outcome; you need to control all the possible variables except the one you're testing. There can be dozens of unsuspected angles, even to an apparently simple experiment like making ice cubes.

Jurassic Stark

Why do icebergs and ice cubes float? Aren't solids generally heavier than liquids?

Generally, yes. But water is an exception. As trivial as this question may sound, the answer is of life-and-death importance. If ice didn't float on water, we might not even be here to wonder why.

Let's see what would happen if ice sank in liquid water. In prehistoric times, whenever the weather got cold enough to freeze the surface of a lake, pond, or river, the ice would immediately have sunk to the bottom. Subsequent warm weather might not have melted it all, because it would be insulated by all the water above it. The next freeze would deposit another layer of ice on the bottom, and so on. Before long, much of the water on Earth, except for an equatorial band where it never freezes, would be frozen solid from the bottom up, and there might not be enough time in the warmer seasons to melt it all the way down. The primitive sea creatures from which we evolved might never have had a chance to develop. Dinosaurs would never have existed. The world would be stark and barren of life.

The floating of solid water (ice) on liquid water is so familiar to us that we don't realize that it's really an unusual phenomenon. When

most other liquids freeze, the solid form is denser, "heavier," than the liquid. That's just what we'd expect, because in solids, the molecules are packed together more tightly and rigidly than they are in free-flowing liquids, so naturally the solids will be heavier and will sink. Try it with a liquid that freezes at a conveniently comfortable temperature: paraffin wax.

TRY IT   *Add a piece of solid wax to some melted wax, and watch it sink. You'd get the same result with molten vs. solid metals, oils, alcohols, and so on. But perform the scientific experiment of placing an ice cube in a glass of water, and you'll get the opposite result. The ice cube floats.*

The reason lies in the unique way in which water molecules are connected to each other in a piece of ice. They are connected by bridges (*hydrogen bonds*) between the water molecules. But consider what a bridge does. Brooklynites might say that the Brooklyn Bridge joins Brooklyn to Manhattan, but Manhattanites might insist that it *separates* Brooklyn from Manhattan, keeping them a bridge-length apart. Well, they're both right, and that's also what hydrogen bonds do to the water molecules in ice: they join the molecules together, but they also hold them a certain distance apart. So instead of crowding together as tightly as the molecules in other solids do, the water molecules form a sort of openwork lattice. The molecules are now on average farther apart in the ice than they were in the liquid, so the ice is taking up more space; it's less dense and will float in liquid water. A given amount of water occupies about 9 percent more space in the form of ice than it does in the form of liquid.

TRY IT   *Look carefully at the ice in your freezer's ice-cube tray. You'll notice that they have little mountain peaks. In freezing, they had to expand, and being restricted on sides and bottom, the only direction they could go was up.*

If freezing water is confined so that it can't expand, it will burst the strongest container in the attempt. That's why a water pipe or an automobile engine can crack when the water inside freezes.

The bridges in ice don't form all at once at the instant of freezing. As we start cooling water down from room temperature, it gets denser and denser, just like any other liquid, because the molecules are slowing down and don't require as much elbow room. Most other liquids just keep getting denser and denser until they freeze, and the solid will be densest of all. But not dear old aqua.

Nitpicker's Corner. In real-life bodies of fresh water, temperature fluctuations, winds, water currents, and other mixing phenomena will mess up these tidy arguments about neat layers of water temperatures. All other things being equal, the principles we've outlined above will prevail. (But according to *Wolke's Law of Pervasive Perversity*, all other things are never equal.)

In the oceans, however, it's a slightly different ball game. Because of all the salt it contains, seawater doesn't have a maximum density at 39°F. As its temperature goes down, it just keeps getting denser and denser and keeps on sinking, all the way down to its freezing temperature. In order for ice to form on the ocean surface, all the water first has to get down to the freezing point. That happens only during a long, hard winter near the North or South Pole.

Bar Bet. Show me a freshwater pond with a layer of ice on the surface and I'll tell you the exact temperature of the water at the bottom, without a thermometer.

Water gets denser and denser only down to a point. When it is cooled to 39.16°F or 3.98 (let's call it 4)°C, it starts going the other way—getting less dense as you cool it; that's because some of those bridges are beginning to form. Finally, at 32°F (0°C), all the rest of the bridges snap into place, the water freezes into ice, and the density falls suddenly to a value that's lowest of all. That's why ice will float on water of any temperature: it's the lightest form of all.

The fact that water has its maximum density at about 39°F (4°C) has further significant consequences for living things. When cold weather cools the surface of a freshwater lake, the water at the surface gets denser and sinks. New water takes its place, gets cooled and sinks. This goes on until all the water in the lake has had a chance to be cooled to its maximum possible density—at 39°F (4°C)—and sink. Only then can the surface water be cooled down those last few degrees to form ice. By the time a surface layer of ice can form on a lake, then, all the water in the lake will already be at a temperature of 39°F (4°C). No matter how cold the weather then gets, any water that gets colder than 39°F stays at the top (because it's lighter), and the fish below can never get any colder and freeze. That's another reason that water's peculiarities are responsible for permitting the existence of life on Earth.

On the Level

How does water "seek its own level"? I mean, how does one part of the water know where the levels of all the other parts are, no matter how distant?

It requires no psychic powers. Just gravity.

"Water seeks its own level" is a tortured phrase that was probably uttered in classic Greek by a philosopher some two thousand years ago, and school teachers have been parroting it ever since. In plain, modern language, what it means is that water will lie flat whenever it can. (But who wouldn't?)

If a body of water—anywhere from a bucket to a bathtub to an ocean—is left undisturbed, it will quickly settle down into a perfectly flat surface, no matter how wavy it was when it started out. It will find the mathematically exact compromise level, averaging out the highs and lows as accurately as any army of surveyors with transits could do. But how indeed does a "hill" know that it must fall, and a "valley" know that it must rise?

It all happens because water (and other liquids) are incompressible. You can't force a liquid to occupy a smaller amount of space by pushing on it, the way you can with a gas. The reason is that the molecules of liquids are already as close together as they can get, and no amount of pressure (within reason) will make them crowd together any closer.

What goes for pushes also goes for pulls. Suppose there's a "hill" on the water's surface. Gravity is trying to pull it down, but its molecules can't oblige by packing down any more compactly; all they can do is spread out sideways into the lower-altitude surrounding territory. The result is that the hill has disappeared and a valley has been filled.

Of course, gravity is also pulling down on the valley water, but it is already as low as it can get. In order to dig itself any deeper it would have to send the excavated water somewhere else: uphill. And that, of course, would be contrary to the force of gravity.

A hill of earth rather than of water would behave in the same way if its molecules could flow past one another as easily as water's can. A hill of sand is an intermediate case; its grains can flow to some extent, so a hill of sand that is too high will "seek its own level" just as water does, although it may never achieve what it seeks. Water is less like a hill of sand than a hill of marbles.

Okay, so you knew all that. But here's a truly startling application of the same principles: the *sight glass*. You've seen them. On the outside of a boiler or other opaque container of water, there is a vertical glass tube, connected to the water inside. You can't see the water level inside the boiler, but you can tell how high it is because it is at exactly the same level as the water in the external glass tube. How does the water in the glass tube know where the water level is inside the boiler?

Well, if the water inside the boiler were temporarily higher than the water in the sight glass, it would level itself down, just as the "hill" did a few paragraphs back. In this case, though, the excess hill-water has no valleys to flow into; it has no place to go except into the glass tube. Result? The tube level goes up and the boiler level goes down. The flow will stop when they reach the same level. Same thing the other way around: if the level in the glass tube were temporarily higher than the level inside the boiler. Either way, they come to exactly the same level.

Clever, these boiler engineers.

TRY IT   *Does your kitchen have one of those plastic-cup gravy separators that look like miniature watering cans? The kind that lets you pour off the juices from the bottom, leaving the top layer of fat behind? It's a great substitute model for a boiler and sight glass. Put some water in it and notice that no matter what the water level is inside the cup (the "boiler") and no matter how you tilt it, the water finds exactly the same level in the transparent spout (the "sight glass").*

Really Haute Cuisine

Why does a boiled egg cook faster in New York City than it does in Mexico City?

It would be great fun if we could attribute the difference to the Big Apple's hustle or to Mexico's *mañana* attitude. Unfortunately, we can't. The difference doesn't even have anything to do with the eggs. It's the water. But it's not what you think.

When it's boiling, water in New York is a little hotter than water in Mexico City. And hotter water will get an egg to a given state of done-ness in a shorter time.

A little thought will show that the biggest difference between New York and Mexico City, apart from the relative difficulty of finding a good corned beef sandwich, is the altitude. The average stove in Mexico City is 7,347 feet higher than one in New York City. And the higher the altitude, the lower the temperature at which water boils. How much lower? If pure water boils at 212°F (100°C) in New York City, it will boil at only 199°F (93°C) in Mexico City. Not a huge difference, but your exemplary three-minute New York City egg will certainly take longer to create in Mexico City.

The reason is simple, once you realize what boiling is: it is when water molecules get hot-and-bothered enough to break away from their brethren in the pot, gather together into rising bubbles of vapor, and fly off into the air as steam. In order to escape, water molecules must have enough energy—that is, they've got to be at a high enough temperature, to overcome two separate forces: (a) they've got to break apart the stickiness (hydrogen bonds) that holds them together in the liquid, and (b) they've got to overcome the pressure that the atmosphere is applying to the surface of the water. That pressure is caused by air molecules that are continually bombarding the surface of the water like a barrage of ricocheting hailstones. The sum-total force of those collisions is transmitted through the water to every molecule within it. Molecules at the surface can just fly off into the huge spaces between the air molecules, but those in the interior of the water must overcome this sum-total pressure in order to get out.

The stickiness of liquid water molecules to each other is the same, of course, whether they're part of a Manhattan or a margarita. But atmospheric pressure is another story. In Mexico City, the air is only 76 percent as dense as it is at sea level. That means that only about three-quarters as many air molecules are bombarding the surface of the

You didn't ask, but . . .

Does that mean that we could make water boil hotter if we artificially increased the pressure on it?

Absolutely. That's precisely what a pressure cooker does. Let's clamp onto a cooking pot a tight-fitting lid with only a small hole for escaping steam. Then we'll place a weight on top of that hole to keep a certain, calculated amount of the steam pressure in, instead of letting it escape freely into the atmosphere. Or else we'll use some sort of pressure regulator to fix the pressure at a predetermined value. The pressure of the "atmosphere" inside the pot will then be maintained at that higher value.

At a typical pressure-cooker pressure of 10 pounds per square inch (0.70 kilogram per square centimeter) above normal atmospheric pressure, the boiling temperature—and hence the temperature of the steam inside—is 240°F (115°C). That's hot enough to make short work of any otherwise long-simmering dish, such as a stew. Moreover, the space inside a pressure cooker is filled with steam, which is a much better conductor of heat than air is. Thus, any heat anywhere in the pot will be conducted into the food more efficiently than if the pot were filled with air. This also makes for faster cooking.

water every second. The water molecules are therefore able to muscle their way upward and boil off without having to have quite so much energy: that is, without having to get quite so hot.

An extreme: The highest point above sea level on this planet is Mount Everest, at 29,028 feet (8,848 meters). At this altitude, the atmospheric pressure is only 31 percent of what it is at sea level, and the boiling temperature of water is only 158°F (70°C). That's not hot enough to cook much of anything, no matter how hungry you may have gotten while climbing.

Teapot in a Tempest

If the boiling temperature of water depends on altitude because the atmospheric pressure differs, won't it also depend on the weather?

According to weather reports, the atmospheric pressure is always changing, even in a single location.

Right you are. But the weather has only a small effect on the boiling temperature of water.

When people go around saying that water boils at 212°F (100°C) at sea level, they're speaking rather loosely. The standard definition of the boiling temperature of pure water says nothing about sea level. It is defined in terms of a specific atmospheric pressure: 29.92 inches (760 millimeters) of mercury, which is a typical, but hardly guaranteed, value for sea-level locations. Every TV weather fan knows that the air pressure changes as the weather changes, whether you live by the seaside or anywhere else. So the temperature of boiling water will indeed depend on whatever the weather conditions happen to be at the time.

Quite arbitrarily, scientists have chosen exactly 760 millimeters of mercury, as the standard pressure they call one atmosphere. (That strange-looking number, 29.92 inches of mercury, is simply a quirk of conversion from millimeters to inches.) The boiling temperature at that standard pressure is called the normal boiling temperature or the normal boiling point. That's what's 212°F or 100°C really is.

While knowing these facts might impress your friends, the effect of atmospheric pressure on the boiling temperature of water isn't big enough to worry about. Even if you were brewing a cup of tea while sitting smack in the eye of a hurricane, where the pressure might drop as low as 28 inches or 710 millimeters of mercury (the world's record low is 25.91 inches or 658 millimeters), the boiling temperature would only go down to 208°F (98°C). It's comforting to know that your tea would still be hot enough.

Skating on Thin . . . Water?

The human record for running speed is about 23 miles per hour. But for ice skating, it's more than 33 miles per hour. Obviously, sliding on ice must increase one's speed. But why is ice so great for sliding? What makes it so slippery?

Actually, solid ice isn't intrinsically slippery. There's a thin film of liquid-like water on its surface that the skaters are sliding on. (I'll explain later why I say "liquid-like" instead of "liquid.")

Solids in general aren't slippery because their molecules are tied tightly together and can't move around freely. The molecules of liquids, on the other hand, are free to move around like a bunch of ball bearings in a teacup, and you know that if you pour those ball bearings on the floor (DO NOT TRY THIS AT HOME!) you can slip on them. So liquids are generally slipperier than solids.

But what scientists can't quite agree upon is exactly what creates the slippery film on the surface of ice. If it comes from slight melting, what makes the ice melt, even at temperatures well below its normal melting point?

Two explanations, pressure melting and friction melting, have been slugging it out during the more-than-a-century that scientists have been puzzling over this simple, everyday phenomenon.

The pressure-melting camp has maintained that it's the pressure of the ice-skate blade on the ice (or the ski on the snow) that does the melting. There's no doubt that ice will melt if you apply pressure to it, because solid ice is less dense, occupying a bigger volume, than the same weight of liquid water. That's why the ice in your Tom Collins floats in the soda; the molecules in ice are spaced at some distance apart, while the molecules in liquid water are slithering intimately over and under one another.

If enough pressure is applied to an ice surface, its molecules will be forced down into the tighter structure of liquid water. That is, some of the stressed ice will melt. Can that be why skaters can slide on ice? Can the pressure of the skates' blades melt some ice, creating a liquid track to slide on? Well, let's see.

The weight of a 150-pound ice skater on an eighth-inch-wide skate blade creates a pressure of only about 25 psi (pounds per square inch). True, speeding hockey players and figure skaters hit the ice with a lot more pressure than that, but it still isn't enough to make tracks that will stay liquid long enough before refreezing. Moreover, a person who is just standing on ice applies very little pressure through the soles of his or her shoes, yet can slip all too easily.

So let's scrap the pressure-melting theory.

What about frictional heat? Rubbing any two solids together, even a skate blade and ice, is bound to create friction, and friction creates heat. According to the friction camp, this heat can melt a streak of liquid as the skate blade or ski skims along the ice or snow. This theory has attracted the majority of disciples over the years.

But recently, some surprising results have been published by scientists using sophisticated, new experimental techniques and theoretical calculations. It is turning out that the surface of pure ice isn't exactly solid—it is semi-liquid, even without any added heat or pressure. By semi-liquid, I mean that the surface molecules are not bound together

Bar Bet. Clean, dry, cold ice is not slippery.

Don't actually try the test in a bar. The bartender's ice probably isn't cold enough; it will be wet and slippery from the outset.

as tightly as those in the body of the solid are, because they have no fellow molecules on top of them to hold them in place. So they can move around or vibrate a bit. A tiny bit of added frictional heat can then make them almost flow like a liquid.

So is the surface film that we slip on a liquid or a solid, or some intermediate state?

For the moment, all we can do is flip a three-sided coin.

TRY IT   *Using a towel to avoid melting it, pick up an ice cube from your freezer compartment or take out a whole tray of ice. Gently feel the ice by passing a finger across its surface. Don't rub too hard. You'll find that the ice isn't slippery at all until your body heat and/or frictional heat have had a chance to warm up the surface and melt it a little.*

The Hose Knows

It must be some ancestral memory from caveman days, but we all seem to know intuitively that water will put out a fire, and we never question it. Well, why does water put out a fire?

Before we get any further, note well: Water must never be used on an electrical fire or on an oil or grease fire. Reasons: Water conducts electricity and can lead it elsewhere, perhaps to your very own feet. And because water won't mix with oil or grease, it just scrambles it around and spreads the fire.

Fire needs three things to survive: fuel, oxygen, and—at least initially—a temperature high enough to ignite the fuel and get the reaction going. After that, the combustion reaction gives off more than enough heat to keep things going.

Obviously, the first thing to do would be to remove the fuel. Nothing to burn, no fire. But water can't do that, so it attacks the other two essentials: the oxygen and the temperature.

A deluge of water from a bucket or hose can smother the fire as if it were a blanket, simply by blocking out the air. A thin layer of water, even for a short time, can do the trick. No air, no oxygen, no fire.

Water can also lower the temperature of the material that's burning. Every combustible material has a minimum temperature that it has to reach before it will ignite and burn. If the water cools the material below that temperature, voilà! no more burning. Even hot water is well below the temperature at which most things can burn.

A deluge isn't necessary. Water from a sprinkler can put out a fire, even though it leaves a lot of oxygen available between the rain drops. So it must work by lowering the temperature. Remember how cooling it was to run through the lawn sprinkler when you were a kid?

A water sprinkler lowers the temperature in two ways. First, water in fine droplets tends to evaporate quickly, and evaporation is a cooling process. Second, water has a peculiarity that makes it much better than almost any other liquid for dousing fires: it is a heat-sucking glutton.

You didn't ask, but . . .
Why won't wet things burn?

As we were saying, water is a champion heat absorber, without getting very hot in the process. When you put a flame to something that's wet, the water soaks up the heat like a sponge, preventing the object itself from ever getting hot enough to ignite.

TRY IT   *This one may astound you. Put a little water in an unwaxed paper cup (not a foam cup) and figure out a way to prop it up so that you can set a candle under it. (Maybe you can sit it on a wire rack that's bridging two coffee cans.) Place a lighted candle under the bottom of the cup. The cup won't burn, but after a while the water will have gotten hot enough to boil, having stolen the heat from the paper as fast as the candle gave it off. Even when the water boils, it will never get hotter than 212°F or 100°C, which isn't anywhere near hot enough to ignite the paper.*

Water has a huge appetite for heat. A pound of water must absorb 0.252 calories of heat before its temperature will go up by a single degree Fahrenheit. Is that a lot? Well, contrast it with the amount of heat it takes to raise the temperature of a pound of several other liquids by one degree Fahrenheit: mercury requires only 0.0083 calories; benzene, 0.063 calories; and olive oil, 0.118 calories.

The moral is that a little bit of water can take away a lot of the fire's heat before boiling away and abandoning the premises as steam. Water is therefore an extremely effective cooling agent. That's why it is used in automobile cooling systems. Of course, being cheap doesn't hurt.

Why Bars Are So Noisy

Why do ice cubes snap, crackle, and pop when I put them in my drink?

Somebody slipped Rice Krispies into your ice-cube compartment. No? Well then, it really must be the ice.

If you listen with a linguist's ear, you'll find that the ice isn't actually popping, which implies a certain hollowness. But it certainly does snap and, on occasion, crackle.

First, the snap. When you plunge a cold ice cube into a warmer liquid, the water warms up parts of the ice cube, which tends to make those parts expand slightly. This places a stress on the ice crystal, because ice has a very rigid structure and it can't just expand here and there at random. The only way the crystal can relieve these stresses is to crack. That's the snap you hear.

Next, the crackling, which sounds like a rapid series of tiny explosions. And that's exactly what it is. Unless you've made your ice cubes out of boiled water (see below), there was some dissolved air in the water that went into your ice tray—or, if you're one of those most fortunate of human beings, into your automatic icemaker. As the water froze, there was no room for the air in the rigid, solid structure that is ice, so the air was trapped as tiny, isolated bubbles. These bubbles are what make the ice cloudy instead of crystal clear.

Now put that bubble-riddled ice cube into a drink. The water works away at melting the surface of the cube, eating its way in deeper and deeper. As it goes, it encounters air bubbles. When those bubbles were formed, they were filled with freezer-cold air, but now they're being warmed up by the advancing water, and they want to expand. But they can't expand until their imprisoning ice walls have been thinned enough to allow them to break through. When it does, Crack! The bubbles explode their way out. Thousands of those tiny break-outs, happening all over the ice surfaces, make a faint crackling or sizzling noise.

The crackles of icebergs and glaciers as they move south into warmer water can be heard loud and clear by the denizens of Arctic-prowling submarines.

TRY IT

Boil some water for several minutes to get most of the dissolved air out of it. Let it cool, pour it into an ice-cube tray, and freeze it. You'll find that there won't be many, if any, bubbles in the ice cubes. (Compare one with an ordinary ice cube by holding them up to a strong light.) When you put the boiled cubes into a drink they may snap, but they won't crackle. You'll enjoy a relatively quiet drink.

... and That's the Way It Is

We have examined well over a hundred everyday happenings and have learned what makes them happen. But is that how science progresses? Questioning each individual happening and finding a unique explanation for it, only to move on to the next, and the next? Not at all.

There are certain general principles that underlie many of the situations we have discussed. Our repetitions showed how interrelated many of the questions really were. It would have been more logical and much more efficient if I had first explained the general principles and then shown how they applied to the several aspects of your everyday life. But then we wouldn't have had a question-and-explanation book; we would have had a textbook. And that's not what you—or I—wanted. (I've been there, done that.)

Nevertheless, those general principles do exist; scientists call them theories. When a theory has been thoroughly tested and has passed with flying colors, it may achieve the exalted status of a Law of Nature. A Law of Nature is simply an elegant way of saying, "This is the way the world works. We may not know why it works that way, but that's the way things are, like it or not."

You've heard of Newton's Law of Gravitation and perhaps of his Laws of Motion. But you may not have heard of the three Laws of Thermodynamics, the powerful laws that govern changes in energy. Science has come up with many other general descriptions of the way things are. The last chapter of this book invokes these general principles to answer some fundamental questions about energy, gravity, mass, magnetism, and radiation, from seeing in the dark to seeing through lead.

This chapter, and the book itself, ends with the seemingly childish, but most profound question of all: "What makes things happen or not happen?" The Second Law of Thermodynamics will give us the answer.

More Heat Than Light

*I don't understand infrared radiation. How can it be used to see
in the dark? People sometimes call it "light" and sometimes call
it "heat." Which is it?*

Strictly speaking, it's neither. It's not light because we can't see it,
and it's not heat because it contains no substance that is capable of
being hot. I like to call it "heat in transit." We'll see why.

Infrared radiation is nothing more than a certain segment of the
broad spectrum of electromagnetic radiations that are being show-
ered down upon us by the sun. Electromagnetic radiations are waves
of energy, traveling through space at the speed of light. Being pure
energy, they are distinguished from the so-called radiations that are
actually streams of tiny particles, such as some of the radiations that
are emitted by radioactive materials.

Electromagnetic radiations differ from one another only by their
energies. The lowest-energy ones are radio waves and the highest-
energy ones are called gamma rays. In between, we find (going up in
energy) microwaves, infrared radiation, visible light, ultraviolet rays,
and X-rays. Gamma rays come mostly from radioactive materials.
Radio waves, microwaves, and X-rays, we can manufacture ourselves.
The rest of this spectrum, this spread of electromagnetic radiation
energies, is provided in abundance by Old Sol.

In order to observe any of these radiations, we must have the right
sort of instrument, attuned to the exact radiation energy we want to
detect. For a certain very small part of the sun's spectrum, we have
a marvelous instrument called the human eye. Not surprisingly, the
part of the spectrum that this instrument is sensitive to is called visible
light. For radio waves and microwaves, we need an antenna to collect
them and electronic circuits to convert them into something that we
can see or hear. For X-rays and gamma rays, we need instruments like
Geiger counters and other paraphernalia that nuclear physicists use.

Infrared radiation (infra-red means "below the red" in energy) is just
outside the energy region that human eyes can detect. That's why we
can't properly call it light. We must detect it by its effect on stuff and
things. Radiations of various energies have different kinds of effects
when they hit matter: when they strike the surface of any substance. In
general, there are three possibilities: the radiation can bounce off (be
reflected), it can be *absorbed*, or it can *penetrate* for a certain distance.

Visible light is reflected by most substances, while X-rays generally
penetrate. But infrared radiation has just the right amount of energy

to be absorbed by the molecules of a wide variety of substances. When a molecule absorbs energy, it becomes, of course, more energetic. It jiggles, rotates, flaps its atoms, and tumbles around more energetically than it did before. And a more energetic molecule is a hotter molecule. So when infrared radiation shines on something, it makes that thing warmer. The radiation itself isn't "heat" until it reaches some kind of substance and gets absorbed. That's why I call it "heat in transit."

You're most likely to see infrared radiation being put to use in two common applications: heat lamps and infrared photography.

Heat lamps are used in restaurants in an attempt to keep your food warm from the time the dish is assembled in the kitchen until your server returns from what appears to be an extended vacation. The lamps are designed to put out most of their "light" in the infrared region of the spectrum, although some of it spills over into visible red light. You can see the glow.

Infrared photography—photography "in the dark," meaning without visible light—is based on the fact that as warm objects lose their heat, they radiate some of it in the form of infrared radiation. This radiation can be detected by special photographic films or by phosphorescent screens. The warm objects are thus rendered visible to people. People themselves are warm, infrared-emitting objects, and are often the targets of those see-in-the-dark snooper devices that work on this principle. Infrared detectors in a helicopter can even find a criminal suspect hiding in the woods by his/her infrared emissions.

Don't Throw Away That Lead-lined Blouse, Lois

Why can't Superman see through lead with his X-ray vision?

He could if he really tried. It's just that his inventors, Jerry Siegel and Joe Shuster, told him that he can see through anything but lead, and like any good cartoon character, he faithfully obeys his creators.

Siegel and Shuster's idea seems to have been that X-rays can't penetrate lead. Otherwise (the reasoning goes), why do X-ray technicians hide behind a lead-lined wall when they zap you? Why do dentists drape you with a lead apron when they shoot your teeth?

Lead is indeed used as radiation shielding throughout the world of nuclear research and technology. But the truth is that there's nothing special about lead at all. It simply does the job more cheaply than other materials.

X-rays are just one kind of electromagnetic radiation: pure energy that zips through space with the speed of light. Other kinds of electromagnetic radiation that are more familiar outside the doctor's office are light itself, the microwaves that cook your food, and the radio waves that carry all that trash to our radios and televisions. All of these energy waves are vibrating up and down and sideways as they fly along. In fact, their energy *consists of* these vibrations: a higher *frequency* of vibration—more vibrations per second—indicates a higher radiation energy. They line up this way in order of increasing energy: AM radio, shortwave radio, television and FM radio, radar, microwaves, light (both visible and invisible to humans), X-rays and gamma rays, the last of which are emitted by radioactive materials and are very penetrating.

Being of such high energy, you might expect (as if you didn't know) that X-rays are quite penetrating radiations. They go through flesh like a bullet through Jell-O. Bones block them just enough to throw diagnostic shadows on a photographic plate or film, the earliest X-ray detectors. Today, there are a wide variety of sensitive digital and electronic detectors that substantially reduce the dosages that patients must be subjected to.

X-rays and gamma rays are *ionizing radiations*. That is, as they plow through atoms of flesh, bone, or anything else, they knock out electrons, leaving behind *ions*—atoms that are missing some of their electrons. And without going into detail, let it be said that atoms that are playing without a full deck of electrons are—to mix a metaphor— loose cannons in the chemical warfare of life. They can disrupt our body chemistry in strange and unhealthful ways. That's why we want to shield ourselves from X-rays and other ionizing radiations, such as those that come from radioactivity.

What, then, shall we use to stop X-rays? Anything that offers lots of atoms with lots of electrons to be knocked out, because each time an X-ray *photon* (an atom's worth of energy) dislodges an electron from an atom, it costs it some of its energy. So the more atoms with more electrons we can put in their way, the sooner the rays will lose all of their energy and stop. The best X-ray stopper or shield, then, is whatever substance has the largest number of electrons per atom and is the most densely packed, with the largest number of atoms per cubic inch.

Uranium would be just dandy. It has 92 electrons per atom and is 19 times as dense as water. Gold would be great, too: 79 electrons per atom and just as dense as uranium. And then there is platinum: 78 electrons per atom and 21 times denser than water. But alas! These substances are all too expensive. And who wants to escape X-rays by hiding behind a wall of radioactive uranium, anyway?

So it all comes down to how many electrons per cubic inch you can buy for a buck. Lead fits the bill better than any other material; it has 82 electrons per atom, is 11.35 times denser than water, and costs only about a dollar a pound.

But some X-rays will always get through a sheet of lead or anything else, no matter how thick; it's just that the thicker the layer, the fewer will get through. In theory, a beam of X-rays can never be stopped completely by any thickness of any material. We can only reduce the beam to a relatively harmless level.

Of course, you can use an even cheaper, if less effective, X-ray stopper than lead; you'll just need more of it. A thick concrete wall, for example, will do the same job as a relatively thin sheet of lead, even though concrete isn't anywhere near as good an X-ray absorber, thickness for thickness. If you have lots of room for shielding, you can even use the cheapest of all materials: water. Only 10 electrons per molecule, but with enough of it between you and the X-ray source, you're safe.

Siegel and Shuster may have known all of this, but admitting it would have spoiled a literary gimmick as great as Achilles' heel. So Lois Lane can rest easily after all in her belief that lead-lined clothing foils X-ray vision.

Until old, mild-mannered Clark wises up.

Cool as a . . . Rutabaga

Why are cucumbers "cool as a cucumber"? I've read in both a cookbook and a food magazine that cucumbers are always 20 degrees cooler than their surroundings. What makes that happen?

Twenty degrees, eh? Well, let's just see about that. (I'll assume that the author of that cookbook was talking about Fahrenheit cucumbers, rather than Celsius.)

If cucumbers are always 20 degrees cooler than their surroundings, let's put a cucumber into a barrel with a whole bunch of other cucumbers and wait to see what happens. Will they fight it out, each one trying to be 20 degrees cooler than its neighbors? Have you ever seen a bushel of cucumbers suddenly freeze itself solid for no apparent reason?

Or, how about this: if cucumbers are always 20 degrees cooler than their surroundings, we'll build a big box out of cucumbers and keep all of our wine nice and cool in it, at maybe 55 degrees. And why stop there? Let's build a smaller cucumber box and put it inside the first one, thereby lowering the temperature by another 20 degrees and we'll keep our beer in it at a nice 35 degrees. And no ice required, thank you, because with one more box we can get down to a temperature well below freezing and make our own. With enough boxes-within-boxes, we could build a refrigerator that would freeze hell itself. And all without even having to plug it in.

We have just violated the most basic law of physics: the *First Law of Thermodynamics*, more familiarly known as the *Law of Conservation of Energy*. For here we have a substance—cucumber flesh—that must be constantly shooting off heat energy into its surroundings. That's the only way an object would be able to stay cool: by constantly throwing off any heat that might flow naturally into it from nearby objects. Since heat is energy, the cucumber flesh is, in effect, an inexhaustible fountain of energy. Free of charge. No need to burn coal or oil or to put up with the problems of nuclear energy. Why, we can use cucumber energy to generate electricity, to propel smog-free automobiles, to irrigate the deserts to grow more and more cucumbers! Why, we can . . .

The only thing we can't do is stop people from putting silly things in books. And of course, the totally fabricated number of 20 degrees

is irrelevant. Automatically cold cucumbers—or automatically cold anything else—simply can't exist. Nothing can permanently maintain even a slightly different temperature—colder or hotter than its surroundings—unless we supply or remove energy to or from somewhere else. That's why we have to plug in our kitchen appliances; we use electrical energy from the local power plant to pump heat energy out of our refrigerators and into our ovens.

But you say you picked up a cucumber that hasn't been in the refrigerator and placed it against your forehead and it really did feel cool? It most certainly did. But that's because the cuke is cooler than your 99-degree skin, not because it's cooler than the 70-degree room.

TRY IT   *Leave an unrefrigerated cucumber and a potato in the same location for a couple of hours. Cut them and hold the cut surfaces against your forehead. They'll feel equally cool. Prove that they're really the same temperature by thrusting a meat thermometer into each.*

Except for variations due to such things as air currents and sunshine streaming through a window, every object in a room is the same temperature. Unless you've turned the thermostat up to your body temperature, they'll all feel cool compared with your skin.

When any two objects are in contact, heat will flow spontaneously from the warmer one to the cooler one. So when the cuke—or any other room-temperature object—sucks out some of your hard-earned forehead heat, you feel the loss as a cool sensation. Cucumber and potato flesh will feel especially cool because they're wet; not only will the water cool you by evaporation, but it makes the vegetable flesh a pretty good conductor of heat that will continue sucking heat out of your skin until it warms up to your skin's temperature.

Scientifically speaking, of course, there's no such thing as coldness; there are only various degrees of heat. The words "cool" and "cold" are mere linguistic conveniences. And so is the expression "cool as a cucumber." It's so much more fun to say than "cool as a rutabaga."

Bar Bet. A cucumber isn't a bit cooler than a potato.

Einstein, Schmeinstein. What's It To Me?

*Everywhere I go (slight exaggeration) I see E=mc², E=mc²,
E=mc². It must be terribly important. I know it has something
to do with the atomic bomb, but what does it really mean to us
folks on the street?*

Frankly, not a helluva lot. But that's not to say that it isn't one of
the most momentous realizations ever to dawn upon the human mind.
It's just that it has to do with things that are happening right under
our noses every day, but that are much too small to notice except when
brought to our attention by that bomb that you mention, which is as-
suredly one of the most effective attention-getting devices of all time.

This most famous of all equations was first put on paper by Albert
Einstein in 1905 as one small part of his theories of relativity. Among
many other things, Einstein discovered that there is an intimate rela-
tionship between mass and energy. Intuitively, we'd love to believe that
energy is energy and objects are objects, period. But Einstein discovered that energy and mass are really two different but interchangeable
manifestations of the same universal stuff, which for want of a better
word we may call *mass-energy*. Einstein's astoundingly simple little
equation is the formula for determining how much energy is equivalent
to how much mass, and *vice versa*.

(For the mathematically unchallenged: If m stands for an amount
of mass and E stands for the equivalent amount of energy, the equation says that you can find that amount of energy simply by multiply-
ing m by the number c^2, the velocity of light, squared. The number
c^2 is incomprehensibly huge, so you can get an enormous amount of
energy from a minute amount of mass.)

The reason that Einstein's equation isn't very relevant in everyday
life (with one major exception that we'll mention) is that all of our
common, daily energy-producing activities, such as metabolizing our
food and burning coal and gasoline, are purely chemical processes, and
in all chemical processes, the amounts of mass that the energy came
from are utterly minuscule.

How minuscule? Well, even if we explode a pound of TNT, which
you will agree is a chemical process that releases a pretty fair amount
of energy, all of that energy comes from the conversion of only half-
a-billionth of a gram (20 trillionths of an ounce) of mass. If we could
weigh the TNT before the explosion and then gather up all the smoke
and gases after the explosion and weigh them, we'd find that they weigh

half-a-billionth of a gram less. That is far, far beneath our notice; we can barely measure such a tiny difference in weight with the world's most sensitive scales. So while Einstein's equation applies without exception to all processes that involve energy—and don't let anybody tell you it doesn't—it is of no consequence whatsoever in our everyday lives.

That goes for all *chemical* processes. *Nuclear* processes, on the other hand, such as the *nuclear fusion* reactions that go on in the sun and the *nuclear fission* reaction that goes on in a nuclear reactor or an atomic bomb, are quite another story. Because virtually all the mass in the world resides in the incredibly dense nuclei of atoms, much greater amounts of energy can be released, atom for atom, in a nuclear process than in a chemical processes. Billions of times greater.

What really makes an atomic bomb the champion of all earthly energy releasers, however, is something called a *chain reaction*. That's a process in which each atom's worth of reaction makes two more reactions, and each of those two makes two more, and each of those four makes two more, and so on, until we have an incredibly huge number of atoms undergoing reaction—all sparked by a single-atom "starter" reaction. When you have an incredibly huge number of atoms reacting within an incredibly short period of time, each one giving off as much energy as a billion ordinary chemical reactions, you've got yourself one hell of an explosion.

Chain reactions aren't all bad. If we control the speed at which a nuclear fission chain reaction multiplies itself, we have a *nuclear reactor*. In a nuclear reactor, the energy is given off gradually enough to generate heat to boil water to make steam to drive generators to make electricity to light the lamp that you may be reading this book by.

That's what's in it for you.

Bar Bet. Mass is converted into energy during ordinary chemical reactions.

Only well-schooled people will bite on this one; you might even hook a chemistry teacher. Chemists are so used to ignoring the tiny mass changes associated with chemical reactions that they believe there simply *are no* mass changes, and that's what ends up being taught in the schools. Clinch your argument by reminding your adversary that Einstein never said, "$E=mc^2$, except in chemistry class."

How Fat Atoms Lose Weight

I can understand that coal and oil must contain energy, because the energy comes out as heat when we burn them. But how do we get energy out of uranium? Does it burn?

If by "burning" you mean *combustion*—a chemical reaction with the oxygen in the air—no. But if you mean do the uranium atoms get used up, yes.

You're right about coal, oil, and uranium containing energy. Actually, every substance contains a certain amount of energy; it is inherent in the unique arrangement of its atoms and how they are held together. If the atoms are held together very tightly, they are in a relatively satisfied state and have low energy. If they are held together only loosely, they have more potential for change; they contain more *potential energy*.

The atoms in nitroglycerin, for example, are very loosely tied together. It is such an unstable substance that it needs only to be bumped a little in order to rearrange its atoms quickly (*very* quickly) into more stable, lower-energy combinations—an assortment of gases. The energy released in the resulting explosion is the difference between the energy of the original nitroglycerin and the energy of the gases into which it has rearranged its atoms.

In general, if we can find a way to rearrange a substance's atoms into a lower-energy grouping, the "lost" energy must come out in some form, usually in the form of heat. When we burn petroleum in air, we're giving its atoms (along with some oxygen atoms) an opportunity to rearrange themselves into lower-energy combinations—carbon dioxide and water. We can then collect the liberated energy as heat. The only reason we can't get energy out of water or stone is that we can't find any lower-energy arrangements of their atoms to transform them into. At least not without expending more energy than we'd get back.

In order to convert themselves into lower-energy atomic combinations, oil, natural gas, and gasoline—all of our usual fuels—must be offered oxygen to react with. Uranium atoms, however, don't need any such help. They can achieve a lower-energy state simply by splitting: dividing their substance into two smaller atoms instead of one big one. The two smaller atoms happen to be tighter, more stable, lower-energy arrangements of subatomic particles (*neutrons* and *protons*) than the original uranium atom was. The consequent decrease in energy is the

Nitpicker's Corner. Why is the uranium nucleus so unstable to begin with, that it is so eager to split in two?

All nuclei are made up of particles called *nucleons*, the generic word for *neutrons* and *protons*. A big nucleus like uranium's is a conglomeration of more than a couple of hundred of these particles, all crowded together into an incredibly tiny space. That's such a large number of objects to hold together that the nucleus's average grip on each one is rather weak. It's like trying to hold a bushel of golf balls together in your arms without benefit of basket.

The nucleus could improve its precarious situation and get a better grip on itself if it could split into two more easily manageable loads—two smaller ("half-bushel") bunches of golf balls—that could be managed more tightly and securely, with more of what scientists call *binding energy*. The two smaller bunches, being better controlled and more stable, would be less likely to scatter; they would have less potential for unruly, energetic behavior: less *potential energy*.

But if energy is mass and mass is energy, then the two smaller nuclei should exhibit their lesser energy by having less mass. As thousands of experiments have shown time and time again, the two "half-bushel" nuclei together do indeed "weigh" less than the single "bushel" nucleus, even though they contain the same number of "golf balls."

If we add up the masses of the two smaller nuclei that the uranium nucleus splits into, we find that they do total about a tenth of one percent less than the mass of the original uranium nucleus. That tenth of a percent of "lost" mass shows up as a huge amount of energy, because according to good old $E=mc^2$, a little bit of mass is equivalent to an enormous amount of energy.

A trifle hard to swallow, perhaps? Remember that these things happen only among submicroscopic particles; we don't see buses spontaneously splitting in two. But if these phenomena weren't real, there wouldn't be such a thing as nuclear energy—or any of the thousands of other nuclear transformations that scientists see in their labs every day; the interconvertibility of mass and energy explains them all beautifully.

Once you've bought into this mass-energy equivalence proposition, you've bought into all of these phenomena as perfectly natural consequences, and they shouldn't be the least bit surprising.

Well, maybe a wee bit.

energy of nuclear fission. Actually, it is only the nucleus of the uranium atom which does the splitting; the rest of the atom (the electrons) just go along for the ride.

But all atoms are not capable of splitting their nuclei to give off energy. Only the very heaviest ones can pull off this trick, called *nuclear fission*. They are so heavy that they're actually a bit wobbly, and they will wobble themselves completely apart—split—at the slightest provocation. A nuclear reactor is essentially a very efficient provocateur. It tickles these wobbly uranium nuclei by lobbing *neutrons*—heavy, uncharged nuclear particles—at them, and that's all the encouragement they need to literally fall apart into more stable arrangements, giving off energy in the process.

Iron-ic Attraction

What makes a magnet attract iron? Why doesn't it attract aluminum or copper, for example?

The only thing that the pole of a magnet will attract is the opposite pole of another magnet. It's exactly the same as with electric charges: the only thing that a positive electric charge will attract is a negative electric charge, and *vice versa*.

While opposite charges or opposite poles do attract each other, there is also an electromagnetic repulsive force: like charges and like poles repel each other. All these actions are manifestations of *electromagnetic force*, one of the four known types of forces in the universe: gravitation, strong nuclear, weak nuclear, and electromagnetic. *Gravitation* is a weak force, but it can operate at very long range; the *strong interaction* is very short-ranged, operating only within atomic nuclei; *the weak force*, responsible for radioactive interactions, is very weak and also short-ranged; and the *electromagnetic force*, both electric and magnetic, is weak but long-ranged.

There is no force at all between an electric charge and something that isn't electrically charged. It's the same with magnets: magnets are attracted only to other magnets. No second magnet, no attraction or repulsion. With magnets, we call the two opposite ends north and south instead of positive and negative, as with electric charges. Historically, we could have named the two opposite kinds of electric charge Male and Female, or Mickey and Minnie, but Benjamin Franklin (yes, *that* Benjamin Franklin) decided to call them positive and negative, and so it has been. The north and south poles of magnets are called *poles* because . . . well, one meaning of the word pole, from the Latin *polus*, meant "one end of the axis of a rotating body."

There are some crossover effects between electricity and magnetism; you can get magnetic attractions from moving charges or electric attractions from moving magnets. We'll consider stationary magnets only.

Iron's atoms are themselves tiny magnets, because their negatively charged electrons—each iron atom has twenty-six of them—behave as if they are spinning like tops as they circle the nucleus, in the same way the Earth spins as it circles the sun. This spinning of electric charges generates one of those "crossover" electric-magnetic phenomena: the spinning electric charges act like magnets. But most of iron's electrons are arranged in such a way that they're all matched up in pairs, one spinning in one direction and the other spinning in the other direction. When electrons pair up in this way, they cancel each other's magnetism, just as two equal-strength bar magnets would: north pole to south pole and south pole to north pole. Net magnetism: nil.

Four of the iron atom's electrons, however, don't have partners, and because they're not paired up they impart a net, uncancelled magnetic effect to the atoms. Iron atoms, therefore, are magnetic and will be attracted to a magnet.

That's all very well, but iron isn't alone, by any means. Dozens of elements—even aluminum and copper—have unpaired electrons in their atoms and are therefore magnetic. Even oxygen atoms have unpaired electrons and are attracted to a magnet. You can't see it happen in the air, of course, but if you pour liquid oxygen onto a strong magnet in a lab, you will see it stick.

This kind of magnetism that comes from unpaired electrons is called *paramagnetism*; it is quite weak, however. It is only about a millionth as strong as the kind of magnetism we usually think of: iron being attracted to a magnet. But you can still observe it at home if you look closely.

TRY IT

Place a carpenter's spirit level on the table and bring a very strong magnet (a rare-earth magnet will do the trick) close to one end of the bubble. Observe closely, using one of the index marks on the tube as a reference point, and if the magnet is strong enough you'll see the air bubble move slightly toward or away from the magnet's pole.

FULL DISCLOSURE: This experiment is a fake. The movement is not caused by the magnet's effect on the air bubble. The magnet is attracting or repelling the *liquid* in the glass tube, which is either *paramagnetic* or *diamagnetic*, two kinds of weak magnetism exhibited by many substances. When the liquid moves one way, the bubble moves the other way, right?

What is different about iron's kind of magnetism (*ferromagnetism*; *ferrum* is Latin for iron) is that in a piece of iron, the atomic magnets need not always be pointing in random, every-which-way directions, like a bunch of compasses in a magnet warehouse.

If we stroke a piece of iron with a magnet, we can drag the iron's atoms into a lined-up arrangement, all pointing their north poles in the same direction and their south poles in the opposite direction. Because of the precise sizes and shapes of the iron atoms, they will then remain in that lined-up arrangement without flopping back. This produces a very strong, additive magnetic effect, millions of times stronger than the magnetism of the individual atoms. The result is that the piece of iron has been "magnetized": it has itself become a magnet and will attract other pieces of iron.

In only three elements, iron, cobalt, and nickel, are the sizes and shapes of the atoms exactly right for them to line up and stay that way. That's why these three metals are the only three ferromagnetic elements. Iron, however, is the strongest.

The World's Biggest BB Gun

If I dropped a BB from the top of the world's tallest building, the 2,722 foot (829.8 m) tall Burj Khalifa in Dubai, United Arab Emirates, and it hit somebody on the head, would it kill him?

No. Pedestrians strolling down Emaar Blvd. at the base of the Burj Khalifa need not fear. Kaffiyeh-hatted or not, they are in little danger from purely scientific experiments such as yours. (We won't deign to discuss water balloons.)

What you undoubtedly have in mind is the acceleration due to gravity—the fact that a falling object will fall faster and faster as time goes by. That is indeed how the physics of falling operates. As an object falls, it is constantly being tugged upon by gravity, so no matter what its speed may be at any given instant, gravity is urging it onward to a still higher speed and it keeps going faster and faster: it *accelerates*. It is the same as if you were pushing a go-cart; so long as you keep pushing, the cart will keep going faster and faster. In a car, we'd call this acceleration "pickup."

Might we not wonder, then, whether if we give our BB enough time to fall, it will eventually be travelling at bullet speed? Or why not the speed of light, for that matter? An actual calculation from the equations of gravity shows that 13 seconds after being released from a height of 830 meters, an object—any object—should hit the ground at a speed of 459 kilometers per hour (285 miles per hour). And surprisingly, as Galileo showed, it doesn't depend on the weight of the object.

But wait. That's assuming that there is nothing at all between the bad BB bomber and his target. But there is. Air. And having to push its way through the air is bound to have a slowing-down effect on a falling object. We now have two opposing forces: the pull of gravity tending to speed the object up and the drag of air tending to slow it down.

Like any opposing forces in nature, these two clever forces work out a mathematically exact compromise. The slow-down from the air cancels out an equal portion of the speed-up from gravity, thereby limiting the ultimate speed that the object will attain no matter how long it falls. It will fall faster and faster only up to a point, and from then on it will fall at a constant speed. That speed is called its *terminal velocity*.

Of course, the air resistance will be different for the various objects that you might consider dropping off a building: much less resistance for a plucked chicken, for example, than for a feathered one. Therefore, the ultimate speed of different objects dropped through the air—their terminal velocities—will be different. If there were no air, their speeds would all be the same after the same amount of time, no matter what their masses.

For a BB, the air resistance is such that its final speed at street level would be quite harmless, even to a bald head. And it would reach that final speed after falling only a few stories, so a scientific expedition to Dubai is entirely unnecessary.

Remember, of course, that it is not only speed, or velocity, that determines the destructive power of a missile; it is its *momentum*. Momentum can be thought of as mass in motion. It is the product of an object's mass and its velocity: $p=mv$. (And don't ask me why physicists use the symbol p to stand for momentum.) Even though a dropped bowling ball's final velocity might not be bullet-like, its effects on a pedestrian could be rather severe because of its mass.

But you probably guessed that.

The Cosmic Boogie

They taught me in chemistry class that all atoms and molecules are in perpetual motion. But then they taught me in physics class that there's no such thing as perpetual motion, and that nothing can keep moving forever without being kicked. (Isaac Newton may not have put it quite that way.) So who's kicking all those atoms and molecules around?

First, a little pedagogical note: Suppose that the Rockettes came on stage one at a time and did solos. You'd agree that the intended effect would be lost, would you not? But that's exactly how school science curricula are designed: the chemistry and physics teachers do solo acts on separate stages. They may not even talk to each other. And there's no course in school called "Putting it All Together." Something should be done about this.

Both of your recollections are correct, of course. The missing link is this: nobody's kicking all those atoms and molecules around now, but they got one helluva kick from *the big bang*, 13.798 ± 0.037 billion years ago. I think it was on a Tuesday.

The movement of atoms and molecules, like the movement of anything else, is a form of energy called kinetic energy (from the Greek *kinema*, meaning "motion"). In the case of atoms and molecules, the kinetic energy is exhibited as an incessant flitting around and crashing into one another, restrained only by bonds, which is how chemists refer to the various kinds of attractions or stickiness between the particles. We call the collective motion of atoms and molecules *heat*.

The fact that all of these particles are in constant motion doesn't mean that any visible-sized chunk of matter—a grain of salt, for example—is going to be bounding about like a Mexican jumping bean. (Those beans have live worms inside, if you must know.) The three billion billion atoms in that grain of salt (yes, I actually calculated it) are oscillating in all possible directions, so they cancel each other out; the salt grain isn't about to suddenly jump off your dining table. A beehive doesn't go galloping across the countryside just because the bees inside are flitting frantically about. (Actually, they're really pretty sedate inside the hive unless you rile them up.)

The big question, then, is where did all the particles in the world get their kinetic energy? Was there one great big, initial shove? Yes, indeed. All the matter in the universe obtained all of its energy at the moment of its creation in the "big bang" that ignited the universe

13.8 billion years ago). And today, billions of years later, every particle in the universe is still quivering.

Not all at the same speed, however. When we add heat energy to a pot of soup on the stove, the soup particles will, on the average, be moving faster. And when we remove heat energy from a bottle of beer by putting it in the refrigerator, the beer particles will, on the average, move more slowly.

You know, of course, that what's going up on the stove and down in the fridge is the *temperature*: a measure of the average kinetic energy of the particles in a sample of matter, whether it is soup, beer, a human being, or a star.

Now we obviously can't climb into a pot of soup with a stopwatch and clock every one of its jillions of particles and average the speeds together to get its temperature. So a man named Daniel Gabriel Fahrenheit had to invent a gadget called a *thermometer*. The thermometer contains a shiny, highly visible liquid in it—mercury—that expands up a glass capillary tube when the temperature goes up and contracts down the tube when the temperature goes down. It expands because the particles whose average kinetic energy we're measuring are clanging up against the glass tube, whose particles in turn are knocked up against the mercury's particles, who then become faster-moving than they were before, and therefore command more elbow room.

Thus, all the atoms and molecules in the universe are still shivering with primordial, universal energy. And energy is all there is; it's the one and only currency in the universe. It can be converted from one form to another, just as money can be converted between the currencies of different nations. It can be lost by one body and gained by another, just as money can be transferred in a financial transaction. It can even be converted into mass, just as money can be converted into goods. The only things it can't do is be created (the mint went out of business right after the big bang) or be destroyed. We obtained a certain amount of it in the big bang, and we've been living on our budget ever since, in the form of heat and all the other forms that energy can be converted into.

In case you think that the sun is continually manufacturing new energy and sending it down to us as heat and light, think again. The sun and stars are just converting into these forms of energy some of the stash of energy they already possess in the form of mass. Nothing new is being generated.

But won't all this ever run down? Won't the cosmic battery, charged up billions of years ago, ever die?

Bar Bet. There *is* such a thing as perpetual motion—at least, motion that has been going on for billions of years without any outside help.

There is every reason to believe that it will. All the energy in the universe is gradually but relentlessly turning into something else: entropy, or disorder—total chaos. But don't worry too much about it. Long before that happens—about six billion years from now, as a matter of fact—the sun will have died. Comforting, what?

Mad About Metrics

Suddenly, not too long ago, American soda pop and liquor began appearing in liter bottles, not quarts or fifths. Was this the first sneaky shot in a coming metric revolution—if we ever decide to join the world? Obstinate as we are, many Americans ask, "Do we really have to switch over to a whole new system of measurements? What's wrong with the system we have now?"

Among all the nations of the world, only four great powers—Brunei, Myanmar, Yemen, and the United States of America—have not yet adopted the metric system of measurement, better known as the International System or the SI (for *Le Système international d'unités* in French). Is it possible that the rest of the world is onto something that has thus far eluded us four countries?

If in the brave new world of metric, you would own a scale that weighs everything out in grams, rather than in pounds and ounces, why would you care how big a gram is? Just measure out 160 of them, 3 of them, 4½ of them and so on, whatever the heck they are. Do you really know what an "ounce" is? All you know is that it's a certain amount of stuff that someone back in the fourteenth century thought would be a convenient amount of weight—or something. Ever since, we must constantly wrestle with three kinds of ounces: fluid, avoirdupois, and troy. They don't even measure the same thing; two of them measure weight and one measures volume.

How about length? Suppose you have a board that measures seven feet, nine and five-eighths inches, and you need to cut it into three equal lengths. Well, good luck with the calculation in English units. (The answer, which you can arrive at in something less than four hours, is two feet, seven and seven thirty-seconds of an inch, more or less.) In the brave new metric world, you would measure the board with a meter stick and find that it is 238 centimeters long. One-third of that is 79.3 centimeters. End of problem.

Note that you didn't have to know or care that there are 2.54 centimeters in an inch. It's only the *conversions* between English and metric units—the old and the new—that can cause trouble. After we're into metric up to our necks, conversions will be irrelevant. An inch? Oh, yeah. That was an ancient unit of measurement, like a cubit.

There's no doubt that it's going to be terribly troublesome to convert everything in the US, from recipes to road maps, not to mention all of our industrial production facilities, to the metric system. Nobody

---

You didn't ask, but . . .

Some US weather reports are already giving temperatures in °C, but the conversion formulas they gave us in school are complicated and impossible to remember. Is there an easy way to convert Celsius to Fahrenheit?

Yes, there is a much simpler way, and it's a shame they don't teach it in school. Once those complicated formulas with all their parentheses and 32s got into a textbook somewhere, they seem to have taken on a life of their own.

Here's the simple method:

To convert a Celsius temperature to Fahrenheit, just add 40, multiply by 1.8, and subtract 40.

That's all there is to it.

For example, to convert 100°C, we'll add 40 to get 140, multiply by 1.8 to get 252, and then subtract 40 to get 212. What do you know! That's the boiling temperature of water: 100°C equals 212°F.

The great thing about this method is that it works in both directions, to wit:

To convert a Fahrenheit temperature to Celsius, just add 40, divide by 1.8, and subtract 40.

Example: to convert 32°F to Celsius, we'll add 40 to get 72, divide by 1.8 to get 40, and then subtract 40 to get—what do you know!— zero. That's the freezing temperature of water: 32°F equals 0°C.

All you have to remember is whether to multiply or divide by 1.8. Hint: Fahrenheit temperatures are always bigger numbers than Celsius. So when you're going toward Fahrenheit, you multiply.

Nitpicker's Corner. This method works because of the ways that Fahrenheit and Herr Celsius set up their temperature scales. It turns out purely by accident—aided by the laws of arithmetic, not physics—that 40 degrees below zero (-40°F and -40°C) represent exactly the same temperature. So adding 40 puts them on the same basis, so to speak. Then all we have to do is correct for the different sizes of the degrees (a Celsius degree is exactly 1.8 times as big as a Fahrenheit degree), and finally we remove the artificial 40 that we added.

A more detailed proof of why this method works would constitute a smaller nit than we care to pick at the moment. But trust me: it always works, and it is exact.

argues with that. But that's the wrong reason for resisting the metric system. Don't we now have to perform ridiculously difficult conversions every day in the English system? Twelve inches in a foot, three feet in a yard, 1,760 yards in a mile, 16 avoirdupois ounces in a pound, 16 fluid ounces in a pint, two pints in a quart, four quarts in a gallon, and so on. Not to mention wrestling with pecks, bushels, barrels, fathoms, knots, and literally hundreds of other crazy units.

In the metric system, there is only one basic unit for each type of measurement. And the only conversion numbers you'll need are 10, 100, and 1,000, not three, four, 12, 16, or 5,280. There are 100 centimeters in a meter, 1,000 meters in a kilometer, 1,000 grams in a kilogram, and so on. Using metrics is simplicity itself, as evidenced by the fact that every schoolchild throughout 94 percent of the world's population has no trouble at all with it.

Once our awkward transition period is over, life will be beautiful. But the longer we wait, the tougher the transition is going to be.

Amen.

Weighty Matters

Why is helium lighter than air? For that matter, why is anything lighter or heavier than anything else?

Everything is made of atoms or molecules. But it's not simply that some atoms are heavier or lighter than others, although that's a big part of it. It is also how tightly the particles are packed together.

Lead is *denser* than—heavier than the same volume of—water. That's mostly because lead atoms are indeed more than eleven times heavier than water molecules. But even if the atoms and molecules were the same weight (*mass*), there could be a difference in density because of the packing. For example, liquid water is denser than solid water (ice), even though they're both made of the same water molecules. But in the liquid, the molecules are packed together more tightly than they are in the solid. So when somebody claims that one substance is denser than another because its particles are heavier, they're not telling the whole story.

Gases are a whole different ball game from liquids and solids, though, because ideally, they don't pack together at all; their molecules are flying around in space, completely free of one another. At the same pressure, all gas molecules will be packed together (actually, *not packed*) to exactly the same extent—that is, they'll be separated by the same average distance, whether they're helium atoms or air molecules.

Thus, packing has nothing at all to do with which gas is denser. Helium is about one-seventh as dense as air at the same pressure, simply because its atoms weigh one-seventh as much as the average air molecule.

Just what you thought, right? But perhaps for the wrong reasons.

Hey! Whose DNA Is This?

What are all those little black smudges that the experts keep waving around in courtrooms as "DNA fingerprints?" Are they the DNA itself?

No. Those ladders of fuzzy black dashes are just a way of making visible to jurors and other devotees of biochemical science certain things that are too small to see, even with a microscope. They're the end result of a number of laboratory manipulations that are never explained in the courtroom. But before we describe them, will the real DNA please stand up?

DNA is the most intricate and awe-inspiring substance on Earth, but it is not too hard to understand if we stay away from the big words and stop just this side of more-than-you-want-to-know.

Suppose that you are Mother Nature, and you want to set up a general scheme of life that would work for all living things, both plant and animal. The biggest problem you face is how to get from one generation to another. After all, manufacturing one exquisite rose, cockroach, or horse, no matter how difficult that may be, isn't going to get you very far unless you give them the power to make more roses, cockroaches, and horses.

How, then, can a rose beget a rose? How can a horse inform its offspring that it should be a horse, rather than a blade of grass or a cockroach, having four legs instead of six, no chlorophyll or antennae, and so on, and so on, and so on, for thousands upon thousands of characteristics? There are an enormous number of explicit specifications that must be prescribed and carried out, to ensure that each succeeding generation follows the same pattern.

How has Mother Nature arranged to record and play back, time after time, without benefit of a computer, the immense amount of complex information that, taken all together, says "horse"?

Answer: she writes it all down on strips of a remarkable substance called DNA, as if on strips of recording tape.

DNA is an abbreviation for the chemical name *deoxyribonucleic acid*. This huge chemical polymer is made up of certain specific clusters of atoms, lined up like rungs spanning a pair of side-by-side ladders that are twisted into interlocking spirals, or helices. This complex assemblage is then coiled up into compact little packages and tucked into the nuclei of virtually every cell of every part of every life-form on Earth, from six-ton elephants to one-celled bacteria.

The information on the DNA ribbons is written in a secret code. (You were expecting Aramaic?) The code consists of the exact sequences in which those rung-like clusters of atoms must be positioned along the twisted ladders. If you think of the atom clusters as words, their sequences are sentences. Specific sequences of atom clusters convey specific pieces of information, just as specific sequences of words do in a sentence.

The atom-cluster "words" are compounds that chemists call *nucleotides* and the "sentences" are called *genes*. Each gene states an essential bit of information about what the baby horse—or cockroach or human being—shall or shall not be, with respect to a specific characteristic. Most remarkably, genes even distinguish each individual baby horse from all others. In a single human DNA ribbon, there are so many "words" (three billion), combined into so many gene "sentences" (about 20,000) that except for identical twins, no two individuals among the seven billion people now on Earth—or among all those who have gone before and are yet to come—have exactly the same combination.

Just imagine the odds. If you had a basket with a few billion words in it and you reached in blindfolded and picked out enough words, one by one, to make a (rather long) book of twenty thousand sentences, what do you think are your chances of repeating the process and getting exactly the same collection of words in the same sequence—of getting exactly the same book? In the case of humans, the odds are even more extreme because of historical and geographic isolation: the odds of exactly duplicating a certain black African baby in a Swedish maternity ward are even slimmer than simple mathematics would indicate.

Aha! Then if every human being on Earth has a unique set of genes on his or her DNA ribbons, can we tell what characteristics an individual has by examining his or her DNA? In principle, yes, except that we haven't yet worked out the entire sequence of nucleotides (the complete *genome*) in very many individuals, although there are commercial labs that will do yours or mine for a price. But if DNA is found in every cell in the body, from skin to blood, hair, fingernails, and semen, couldn't we identify the perpetrator of a crime, for example, by matching his or her DNA with the DNA from cells found at the scene? Definitely. And that's what forensic DNA analysis is all about.

How do they do it? They extract the DNA from the cell samples and treat it with enzymes that "grow" the DNA—make repeated "Xerox" copies of it—until they have enough to work with. Other enzymes are then used to cut up the ribbons into various manageable-sized fragments, like cutting up a book into various pages, paragraphs, sentences, and phrases. Then the technicians spread out all the cut-up fragments according to their sizes (I'll tell you how) and compare exactly which arrangements of words show up in both of the samples that they want to compare. "Same fragments" means "same DNA" and same person.

Think about it. If you can cut two books into hundreds of pieces and wind up with even half a dozen identical pages or assortments of paragraphs in exactly the same order, then by golly you've got two copies of the same book. (Or one helluva case of plagiarism.)

Now about those infamous black smudges. Those ladders of thick, black lines are made by the fragments of DNA, spread out along a kind of racetrack according to their sizes by an electrical technique called *gel electrophoresis*. Technicians give the fragments a negative electric charge and allow them to drift slowly toward a positive electric electrode. The smallest, lightest fragments drift fastest and travel farthest, winding up at the top of the ladder when the race ends; heavier fragments lag behind to various degrees. Thus, they are spread out according to their sizes.

The invisibly small groups of separated DNA fragments are made radioactive, so that their radiations will expose spots on a sheet of photographic film, thus visually revealing their final locations on the racetrack. That developed sheet of film, bearing black exposure marks wherever the fragments wound up at the end of the race, is what the scientists compare, thereby comparing the DNA structures of the two samples. The same finish positions at the end of the race indicate the same DNA and therefore the same individual, with odds that can be as high as hundreds of trillions to one.

Of course, there is always a slim chance that the murderer was a horse.

Use It and Lose It

To save energy and resources, we're recycling all sorts of things these days. Can we recycle energy itself?

Absolutely, if by recycling you mean transforming something into a more useful form. We do it all the time. Power plants transform water, coal, or nuclear energy into electricity. In our kitchen toasters we transform electrical energy into heat energy. In our automobile engines we transform chemical energy into motion (*kinetic energy*). The different forms of energy are all interchangeable; all we have to do is invent an appropriate machine to do the job.

But there's a catch—perhaps the biggest single catch in the entire universe: Every time we convert energy, we lose some of its value. That's not just because our gadgets are inefficient or because we're sloppy; it is much more fundamental than that. It's like converting currency in a foreign country; there is a cosmic exchange agent who inevitably takes a little cut out of each transaction. The name of this cosmic exchange agent is the *Second Law of Thermodynamics.*

It's really a good-news/bad-news joke.

First, the good news. It's the *Law of Conservation of Energy*, also known as the *First Law of Thermodynamics*. It says that energy cannot be created or destroyed. It can be changed back and forth from one of its many forms to another—heat, light, chemical, electrical, mass, and so on—but according to the First Law the amount of it must always stay the same; energy never just disappears. The amount of mass-energy in the universe was fixed at the time of its creation. We can never run out of energy.

Great! Then all we have to do is keep converting and reconverting our energy into whichever form we happen to need at the moment—light from a bulb, electricity from a battery, motion from an engine—and keep using it over and over. We'll recycle energy just as we recycle aluminum cans, right?

Unfortunately, wrong. Here's the bad news. The Second Law of Thermodynamics says that every time we convert energy from one form to another, we lose a little of its usefulness. We can't lose any energy itself—the First Law prohibits that—but we lose some of its *ability to do work*. And if you can't do work with it, what good is energy, anyway?

The reason that some work strength is lost is that every time we convert energy from one form into another, some of it winds up as heat energy, whether we want it to or not. About 60 percent of the energy in the coal that they burn down at the local power plant is wasted as heat; only about 40 percent of it winds up as electricity, and much of that is lost on its way to you through those overhead wires. Then, 98 percent of the electrical energy you put into a lightbulb is wasted as heat.

Much of the chemical energy in gasoline goes out the radiator and tailpipe of your car as heat. Even if all these complex operations were 100 percent efficient, some heat would inevitably be lost. Even when water turns a waterwheel, a little bit of the water's energy winds up as frictional heat in the wheel's bearings.

Expecting no heat to be formed at all is like expecting no friction. And expecting no friction would be expecting a machine to keep running forever without ever slowing down: *perpetual motion*. Energy from nowhere. And that's impossible. (See the First Law.) Therefore, wherever energy is being put to work, some heat must be formed.

But heat is still energy, isn't it? Sure it is. Then why can't we just take that heat and put it back to work as usable energy?

Well, here's the real bad news of the Second Law: We can indeed do that, but not completely. *While other forms of energy can be converted 100 percent into heat, heat cannot be converted 100 percent into any other form.* Why? Because heat is a random, disordered motion of molecules. And once your energy is in that chaotic condition, you just can't get a full complement of useful work out of it. Just try to plow a field with a "team" of horses that are running around in all directions.

So little by little, as the world spins on, all forms of energy are relentlessly being converted into irretrievable heat. The world's energy is gradually turning into a useless, chaotic, random motion of particles. As far as energy is concerned, the more we use, the more we lose.

The universe is running down like a cheap battery.

We're on a one-way street, headed into a dead end.

Have a nice day.

This May Make Your Head Spin

*The Earth is rotating. The sun is rotating. Are all the heavenly
bodies rotating, or spinning? If so, what started them?*

The quick answer is: why *shouldn't* they be spinning?

Have you ever played billiards, or pool? One popular game, American pool, starts when one player shoots the cue ball (the white one) forcefully at a tight grouping, usually triangular, of the fifteen other balls. The impact of the cue ball breaks up that neat arrangement and scatters the balls all over the table.

The balls are now rolling, right? Wouldn't you be surprised if they weren't? But how can a ball roll if it isn't rotating in the direction it's moving in? A wheel can't get anywhere if it isn't turning. So all the balls have suddenly begun rotating the moment they were impacted by the cue ball or by one of their fellow target balls. Because the target balls were tightly bunched in contact with one another, the cue ball's energy was swiftly spread over the whole bunch. The "rotation virus" spread instantly, like an epidemic of dizziness.

So does every collision between billiard balls—or planets or stars or galaxies—result in rotations? The odds are overwhelmingly, yes.

In pool, the shooter must align his or her cue stick precisely at the center of the cue ball, that is, along a true diameter of the ball. A little to this side or that side will start the ball spinning off into some unintended direction. The only way to direct the cue ball where you want it is to smack it dead-center. And that isn't easy, because on the surface of the cue ball there is an infinite number of other, wrong, places to hit it.

Now, as to the heavenly bodies.

After the big bang, which is the most widely accepted story of the origin of the universe, everything was a huge, roiling cloud of dust, gases, and energy in the form of radiation and motion (*kinetic energy*). By and by (not a precise scientific term), under the influence of *gravitation, electromagnetic forces*, and *radiation pressure*, the dust and gases began to cluster together to form stars and eventually, galaxies. In this chaotic setting, collisions between bodies were inevitable. But just as with billiard balls, glancing collisions were much more probable than direct, head-on collisions, and glancing collisions make both colliding objects spin.

But the dusts and gases, plus their conglomerations and coagulations, were spinning in all sorts of different directions before they clumped together. Then how were they able to condense into a body, say a star, that rotates as a whole in a single direction?

What I've been calling spin or rotation is what physicists call *angular momentum*. And one fundamental Law of Nature that we've never seen violated is the *Law of conservation of angular momentum*: in any interaction between moving, rotating bodies, the total amount of angular momentum remains constant. For example, if a spinning ball hits a stationary ball of the same mass, the amount of angular momentum in the spinning ball can be shared between the two post-collision balls in many ways, but the total amount of angular momentum in both of them must equal the angular momentum brought in by the spinning one. (The stationary one had no angular momentum.)

Another example: If two same-mass spinning bodies collide, one while spinning clockwise and the other while spinning at the same speed (with the same *angular velocity*) counterclockwise, their angular momenta will cancel each other, and the net angular momentum of the system will be zero. If they remain isolated from all other bodies and forces, then no matter what they do after the collision, their net angular momentum must remain zero.

So when the components of the dust-and-gas cloud clump together to form a star, all their angular momenta average out into whichever spin direction is in excess, and the entire star spins in that direction with that momentum.

And that's how rotations began to pervade the universe, all the way down to our own star and our own little old home planet, which rotates west to east exactly once a day.

Coincidence? No, that's how we defined the day.

Why Does That Happen, Daddy?

This may be a stupid question, but what makes things happen or not happen? I mean, water will flow downhill, but not up. I can put sugar in my coffee, but if I put in too much I can't get it out again. I can burn a match, but I can't unburn it. Is there some cosmic rule that determines what can happen and what can't?

There is no such thing as a stupid question. Actually, this is perhaps the most profound question in all of science. Nevertheless, it does have a fairly simple answer, ever since Josiah Willard Gibbs figured it all out in the late nineteenth century.

The answer is that everywhere in Nature there is a balance between two fundamental qualities: *energy*, which you probably know something about, and *entropy*, which you may not (but soon will). It is this balance alone that determines whether or not something can happen.

Certain things can happen all by themselves, but they can't happen in the opposite direction unless they get some outside help. For example, we could make water go uphill by hauling it or pumping it up. And if we really wanted to, we could get that sugar back out of the coffee by evaporating the water and then chemically separating the sugar from the coffee solids. Unburning a match is quite a bit tougher, but given enough time and equipment, a small army of chemists could probably reconstruct the match out of all the smoke and gases.

The point is that in each of these cases a good deal of meddling— energy input from outside—is required. Left entirely to herself, Mother Nature allows many things to happen spontaneously, all by themselves. But other things will never happen spontaneously, even if we wait, hands-off, until doomsday. Nature's grand bottom line is that if the balance between energy and something called *entropy* is proper, it will happen; if it is not, it won't.

Let's take energy first. Then we'll explain entropy.

In general, everything will try to decrease its energy if it can. At a waterfall, the water gets rid of its pent-up gravitational energy by falling down into a pool. (We can make that cast-off energy turn a water wheel for us on the way down.) But once the water gets down to the pool, it is "energy-dead," at least gravitationally speaking; it can't get back up to the top. A lot of chemical reactions will happen for a similar reason: the chemicals are getting rid of their pent-up energy by spontaneously transforming themselves into different chemicals that have less energy. The burning match is one example.

Thus, other things being equal, Nature's inclination is that everything will lower its energy if it can. That's rule number one.

But decreasing energy is only half the story of what makes things happen. The other half is increasing entropy. Entropy is just a fancy word for the disorder, or randomness, the chaotic, irregular arrangement of things.

At the scrimmage line, football players are lined up in an orderly arrangement; they are not disorderly, and they therefore have low entropy. After the play, however, they may be scattered all over the field in a more disorderly, higher-entropy arrangement. It's the same for the individual particles that make up all substances: the atoms and molecules. At any given time, they can be in an orderly arrangement, or in a highly disordered jumble, or in any kind of arrangement in between. That is, they can have various amounts of entropy, from low to high.

But other things (namely, energy) being equal, Nature's inclination is that everything tends to become more and more disorderly—that is, everything will increase its entropy if it can. That's rule number two. There can be an "unnatural" increase in energy as long as there is a more-than-compensating increase in entropy. Or, there can be an "unnatural" decrease in entropy as long as there is a more-than-compensating decrease in energy.

So the question of whether or not a happening can occur spontaneously in Nature—without any interference from outside—is a question of balance between the energy and entropy rules.

The waterfall? That happens because there is a big energy decrease; there's little entropy difference between the conditions of the water at the top and at the bottom. It's an energy-driven process.

The sugar in the coffee? It dissolves because there's a big entropy increase; sugar molecules swimming around in coffee are much more disorderly than when they were tied neatly together in the sugar crystals. Meanwhile, there is virtually no energy difference between the solid sugar and the dissolved sugar. (The coffee doesn't get hotter or colder when the sugar dissolves, does it?) It's an entropy-driven process.

The burning match? Obviously, there's a big energy decrease; the pent-up chemical energy is released as heat and light. But there is also a huge entropy increase; the billowing smoke and gases are much more disorderly than the compact little match head was. So this reaction is doubly blessed by Nature's rules, and it occurs with great gusto the instant you provide the initiating scratch. It's driven by both energy and entropy.

What if we have a process in which one of the quantities, energy or entropy, goes the "wrong way"? Well, the process can still occur if the other quantity is going the "right" way strongly enough to overcome it. That is, energy can increase as long as there's a big enough entropy increase to counterbalance it; and entropy can decrease as long as there's a big enough energy decrease to counterbalance it. What Gibbs did was to devise and write down an equation for this energy-entropy balance. It happens that if this equation comes out with a negative sign, the process in question is one that Mother Nature permits to occur spontaneously; if it comes out with a positive sign, the process is impossible. Absolutely impossible, unless human beings or something else sidesteps the rules by bringing in some energy from outside.

By using enough energy, we can always overpower Nature's entropy rule that everything tends toward disorderliness. For example, with enough effort we could collect, atom by atom, the ten million tons of dissolved gold that are distributed throughout the Earth's oceans, sitting there just for the taking. But it is dispersed through 324 million cubic miles (1.35 billion cubic kilometers) of ocean in a random, incredibly high-entropy arrangement. The problem is that the energy necessary to collect it would cost a lot more than the value of the gold.

In a fit of fervor over the laws of mechanics, Archimedes (287–212 B.C.) is reputed to have said, "Give me a lever long enough and a place to stand, and I will move the world." If he had known about entropy—and apple pie—he might have added, "Give me enough energy and I will diminish the world's entropy down to apple-pie order."

Buzzwords

Acceleration: The rate at which the velocity of a moving body changes with time.

Acid, base, and salt: Acids and bases are opposite kinds of chemicals that neutralize one another, forming water and a salt. Table salt is the most common salt. Common acids are carbon dioxide, and vinegar. Common bases are ammonia and lye.

Alloy: A metal that has been made by melting together two or more pure metals.

Atom: The smallest unit of an element. There are 118 known kinds of atoms. Atoms are almost always joined together in various combinations to form molecules.

Calorie: An amount of heat energy. As chemists use the word, a calorie is the amount of heat that it takes to raise the temperature of one gram of water by one degree Celsius. The food calorie used by nutritionists, however, is equal to 1,000 of these chemists' calories; chemists would call it a kilocalorie. (No, they will never agree.) In this book, we use the nutritionists' calorie exclusively.

Capillary: A very thin tube, or any very thin space through which a liquid can flow. Water and some other high-surface-tension liquids creep automatically into such thin spaces because their molecules are attracted to the walls of the tube.

Carbohydrates: A family of plant chemicals that includes starches, sugars, and cellulose.

Chemical compound: A pure, definable substance whose molecules are made up of definite types and numbers of atoms.

Condensation: When a vapor cools enough to become a liquid, it is said to condense. Condensation is the reverse of boiling, in which a liquid gets hot enough to turn into a vapor.

Convection: The transmission of heat by movement of a hot substance, such as water or air.

Crosslinking: The process by which adjacent polymer molecules form a bond between them.

Crystal: A solid that is made up of a regular geometric arrangement of particles. The solid reflects this regular internal arrangement by having a regular geometric outer shape.

Density: A measure of how heavy a given volume (bulk) of a substance is. A cubic foot of water, for example, weighs 62.4 pounds; the density of water is therefore 62.4 pounds per cubic foot. In metric units, the density of water is one gram per cubic centimeter. For comparison, the density of lead is 11 grams per cubic centimeter.

Dissolving: When a substance dissolves in water, it seems to disappear, because it actually does come apart; its molecules separate from one another and mix in intimately amongst the molecules of water. This mixture is called a solution.

Electromagnetic radiation: Pure energy in wave form, traveling through space at the speed of light. Known types of electromagnetic energy range from radio waves to microwaves to light (both visible and invisible) to X-rays to gamma rays. Electromagnetic waves have a wavelength and a frequency of vibration; the shorter the wavelength, the higher the frequency and energy.

Electron: A tiny, negatively charged particle. Its native habitat is outside the nucleus—the extremely heavy core—of an atom. Electrons are easily detached from their atoms, and can move about on their own.

Enzyme: A natural catalyst—a substance that speeds up a chemical process without itself being used up or changed in any way. Enzymes in plants and animals make otherwise too-slow life processes take place at a reasonable speed.

Essential mineral: A chemical element that is necessary in the diet for life and health.

Fluorescence: The process in which a substance absorbs energy and re-emits it as light.

Heat: A form of energy, manifested by the motion of the atoms and molecules in a substance.

Heat capacity: The amount of added heat that is required to raise the temperature of a substance by a certain number of degrees. Water, for example, soaks up a great deal of heat before getting much hotter; it has a high heat capacity.

Heat of fusion: The amount of heat it takes to melt a solid, usually expressed as the number of calories needed to melt one kilogram of it.

Humidity: A measure of the water vapor content of the air. Usually expressed as relative humidity, relative to the air's maximum content of water vapor at that temperature.

Hydrogen bond: A weak attraction between water molecules that hinders their free movement around one another. It is responsible for many of the unique properties of water.

Infrared radiation: Electromagnetic radiation of energy just below that of red visible light.

Inorganic or organic compound: Chemists have divided all chemicals into two classes: inorganic and organic. Organic compounds contain carbon atoms in their molecules; inorganic compounds don't. Almost all of the chemicals involved in plant and animal life are organic. "Organic" as applied to foods can mean anything the seller wants it to mean. Often, it means little more than "higher priced."

Ion: An ion is an atom or group of atoms that have an electric charge, obtained by losing some of their electrons or by gaining extra ones. Most minerals exist in water as ions, rather than as uncharged atoms or molecules.

Kinetic energy: Kinetic energy is energy that is in the form of action, or motion. A pitched baseball has obvious kinetic energy. But heat is also a form of kinetic energy, because it consists of the movement of the atoms and molecules in an object, even though the object itself may not be moving.

Molecule: A tiny particle, of which almost all substances are made. Different substances are different because their molecules are different in composition, arrangement, size, or shape. Molecules, in turn, are made of even tinier particles called atoms. Atoms, in their turn, are made of electrons, distributed around a nucleus.

Nucleus (plural, nuclei): The heavy central core of the atom, containing virtually all of the atom's mass or weight. It is thousands of times as heavy as the atom's electrons.

Oxidation: A chemical reaction in which an entity loses one or more electrons.

Phosphor: A substance that emits light when excited, or stimulated.

Photon: An elementary quantity of electromagnetic radiation, analogous to an atom being an elementary quantity of matter.

Polar: A polar substance is made up of molecules whose electrons are more concentrated at one end than at the other, which makes that end of the molecule negatively charged, compared with the other end. Such a molecule responds to electric and magnetic forces, whereas a non-polar molecule would be unaffected. Water molecules are strongly polar, which gives water some unique properties.

Polymer: A substance whose large molecules are made up of many smaller molecules, all tied together. Plastics and proteins are polymers.

Potential energy: Energy that is somehow stored up and waiting to be released to do useful work. Examples: gravitational potential energy (a boulder poised on the rim of a canyon), chemical potential energy (a stick of dynamite), and nuclear energy (a critical collection of uranium atoms).

Pressure: The amount of force that is being applied to each and every unit of an area. Often measured in pounds per square inch.

Protein: A polymer found in plants and animals, whose large molecules have been formed by amino acid molecules condensing together. Amino acids are nitrogen-containing, organic chemical compounds that are essential to human structure and metabolism.

Redox reaction: A chemical reaction in which electrons are being passed from one kind of atom, molecule, or ion to another.

Reduction: A chemical reaction in which an atom or molecule receives one or more electrons.

Soap: A salt made by the reaction of a fatty acid with an alkali.

Spectrum: A display of all the wavelengths of radiation that a given substance emits or absorbs. The sun emits a broad spectrum of radiation, including the visible spectrum: the rainbow of colors that the human eye can see.

Surface tension: Liquid water's surface molecules are attracted more strongly to one another (by hydrogen bonds) than to the air molecules above its surface. The result is an inward force, or tension, at the surface that makes it behave as if it were covered with an elastic membrane.

Temperature: A number that expresses the average kinetic energy, or motion energy, of all the particles in a substance.

Index

ABOUT THE AUTHOR

Robert L. Wolke (PhD, Cornell University) is a scientist turned journalist and author. Currently Professor Emeritus of Chemistry at the University of Pittsburgh, he has long been known for his ability to make science both understandable and enjoyable for liberal arts students. He has also taught at the Universities of Florida, Puerto Rico, and Oriente in Venezuela, and has carried out research in nuclear chemistry at Oak Ridge National Laboratory and the University of Chicago's Enrico Fermi Institute. He is the author of *Impact: Science on Society*, *Chemistry Explained*, the "Einstein series" of everyday science books, and the award-winning food science column "Food 101" for *The Washington Post*, as well as many articles and essays for both national and international magazines and newspapers. He lives, writes, and cooks in Pittsburgh with his wife, Marlene Parrish, a food writer and restaurant consultant.

A CATALOG OF SELECTED
DOVER BOOKS
IN ALL FIELDS OF INTEREST

A CATALOG OF SELECTED DOVER
BOOKS IN ALL FIELDS OF INTEREST

100 BEST-LOVED POEMS, Edited by Philip Smith. "The Passionate Shepherd to His Love," "Shall I compare thee to a summer's day?" "Death, be not proud," "The Raven," "The Road Not Taken," plus works by Blake, Wordsworth, Byron, Shelley, Keats, many others. 96pp. 5³⁄₁₆ x 8¼. 0-486-28553-7

100 SMALL HOUSES OF THE THIRTIES, Brown-Blodgett Company. Exterior photographs and floor plans for 100 charming structures. Illustrations of models accompanied by descriptions of interiors, color schemes, closet space, and other amenities. 200 illustrations. 112pp. 8⅜ x 11. 0-486-44131-8

1000 TURN-OF-THE-CENTURY HOUSES: With Illustrations and Floor Plans, Herbert C. Chivers. Reproduced from a rare edition, this showcase of homes ranges from cottages and bungalows to sprawling mansions. Each house is meticulously illustrated and accompanied by complete floor plans. 256pp. 9⅜ x 12¼. 0-486-45596-3

101 GREAT AMERICAN POEMS, Edited by The American Poetry & Literacy Project. Rich treasury of verse from the 19th and 20th centuries includes works by Edgar Allan Poe, Robert Frost, Walt Whitman, Langston Hughes, Emily Dickinson, T. S. Eliot, other notables. 96pp. 5³⁄₁₆ x 8¼. 0-486-40158-8

101 GREAT SAMURAI PRINTS, Utagawa Kuniyoshi. Kuniyoshi was a master of the warrior woodblock print — and these 18th-century illustrations represent the pinnacle of his craft. Full-color portraits of renowned Japanese samurais pulse with movement, passion, and remarkably fine detail. 112pp. 8⅜ x 11. 0-486-46523-3

ABC OF BALLET, Janet Grosser. Clearly worded, abundantly illustrated little guide defines basic ballet-related terms: arabesque, battement, pas de chat, relevé, sissonne, many others. Pronunciation guide included. Excellent primer. 48pp. 4³⁄₁₆ x 5¾. 0-486-40871-X

ACCESSORIES OF DRESS: An Illustrated Encyclopedia, Katherine Lester and Bess Viola Oerke. Illustrations of hats, veils, wigs, cravats, shawls, shoes, gloves, and other accessories enhance an engaging commentary that reveals the humor and charm of the many-sided story of accessorized apparel. 644 figures and 59 plates. 608pp. 6 ⅛ x 9¼. 0-486-43378-1

ADVENTURES OF HUCKLEBERRY FINN, Mark Twain. Join Huck and Jim as their boyhood adventures along the Mississippi River lead them into a world of excitement, danger, and self-discovery. Humorous narrative, lyrical descriptions of the Mississippi valley, and memorable characters. 224pp. 5³⁄₁₆ x 8¼. 0-486-28061-6

ALICE STARMORE'S BOOK OF FAIR ISLE KNITTING, Alice Starmore. A noted designer from the region of Scotland's Fair Isle explores the history and techniques of this distinctive, stranded-color knitting style and provides copious illustrated instructions for 14 original knitwear designs. 208pp. 8⅜ x 10⅞. 0-486-47218-3

AN ENCYCLOPEDIA OF BATTLES: Accounts of Over 1,560 Battles from 1479 B.C. to the Present, David Eggenberger. Essential details of every major battle in recorded history from the first battle of Megiddo in 1479 B.C. to Grenada in 1984. List of battle maps. 99 illustrations. 544pp. 6½ x 9¼. 0-486-24913-1

ENCYCLOPEDIA OF EMBROIDERY STITCHES, INCLUDING CREWEL, Marion Nichols. Precise explanations and instructions, clearly illustrated, on how to work chain, back, cross, knotted, woven stitches, and many more — 178 in all, including Cable Outline, Whipped Satin, and Eyelet Buttonhole. Over 1400 illustrations. 219pp. 8⅜ x 11¼. 0-486-22929-7

ENTER JEEVES: 15 Early Stories, P. G. Wodehouse. Splendid collection contains first 8 stories featuring Bertie Wooster, the deliciously dim aristocrat and Jeeves, his brainy, imperturbable manservant. Also, the complete Reggie Pepper (Bertie's prototype) series. 288pp. 5⅜ x 8½. 0-486-29717-9

ERIC SLOANE'S AMERICA: Paintings in Oil, Michael Wigley. With a Foreword by Mimi Sloane. Eric Sloane's evocative oils of America's landscape and material culture shimmer with immense historical and nostalgic appeal. This original hardcover collection gathers nearly a hundred of his finest paintings, with subjects ranging from New England to the American Southwest. 128pp. 10⅝ x 9. 0-486-46525-X

ETHAN FROME, Edith Wharton. Classic story of wasted lives, set against a bleak New England background. Superbly delineated characters in a hauntingly grim tale of thwarted love. Considered by many to be Wharton's masterpiece. 96pp. 5$\frac{3}{16}$ x 8 ¼. 0-486-26690-7

THE EVERLASTING MAN, G. K. Chesterton. Chesterton's view of Christianity — as a blend of philosophy and mythology, satisfying intellect and spirit — applies to his brilliant book, which appeals to readers' heads as well as their hearts. 288pp. 5⅜ x 8½. 0-486-46036-3

THE FIELD AND FOREST HANDY BOOK, Daniel Beard. Written by a co-founder of the Boy Scouts, this appealing guide offers illustrated instructions for building kites, birdhouses, boats, igloos, and other fun projects, plus numerous helpful tips for campers. 448pp. 5$\frac{3}{16}$ x 8¼. 0-486-46191-2

FINDING YOUR WAY WITHOUT MAP OR COMPASS, Harold Gatty. Useful, instructive manual shows would-be explorers, hikers, bikers, scouts, sailors, and survivalists how to find their way outdoors by observing animals, weather patterns, shifting sands, and other elements of nature. 288pp. 5⅜ x 8½. 0-486-40613-X

FIRST FRENCH READER: A Beginner's Dual-Language Book, Edited and Translated by Stanley Appelbaum. This anthology introduces 50 legendary writers — Voltaire, Balzac, Baudelaire, Proust, more — through passages from *The Red and the Black, Les Misérables, Madame Bovary,* and other classics. Original French text plus English translation on facing pages. 240pp. 5⅜ x 8½. 0-486-46178-5

FIRST GERMAN READER: A Beginner's Dual-Language Book, Edited by Harry Steinhauer. Specially chosen for their power to evoke German life and culture, these short, simple readings include poems, stories, essays, and anecdotes by Goethe, Hesse, Heine, Schiller, and others. 224pp. 5⅜ x 8½. 0-486-46179-3

FIRST SPANISH READER: A Beginner's Dual-Language Book, Angel Flores. Delightful stories, other material based on works of Don Juan Manuel, Luis Taboada, Ricardo Palma, other noted writers. Complete faithful English translations on facing pages. Exercises. 176pp. 5⅜ x 8½. 0-486-25810-6

THE RED BADGE OF COURAGE, Stephen Crane. Amid the nightmarish chaos of a Civil War battle, a young soldier discovers courage, humility, and, perhaps, wisdom. Uncanny re-creation of actual combat. Enduring landmark of American fiction. 112pp. 5³⁄₁₆ x 8¼. 0-486-26465-3

RELATIVITY SIMPLY EXPLAINED, Martin Gardner. One of the subject's clearest, most entertaining introductions offers lucid explanations of special and general theories of relativity, gravity, and spacetime, models of the universe, and more. 100 illustrations. 224pp. 5⅜ x 8½. 0-486-29315-7

REMBRANDT DRAWINGS: 116 Masterpieces in Original Color, Rembrandt van Rijn. This deluxe hardcover edition features drawings from throughout the Dutch master's prolific career. Informative captions accompany these beautifully reproduced landscapes, biblical vignettes, figure studies, animal sketches, and portraits. 128pp. 8⅜ x 11. 0-486-46149-1

THE ROAD NOT TAKEN AND OTHER POEMS, Robert Frost. A treasury of Frost's most expressive verse. In addition to the title poem: "An Old Man's Winter Night," "In the Home Stretch," "Meeting and Passing," "Putting in the Seed," many more. All complete and unabridged. 64pp. 5³⁄₁₆ x 8¼. 0-486-27550-7

ROMEO AND JULIET, William Shakespeare. Tragic tale of star-crossed lovers, feuding families and timeless passion contains some of Shakespeare's most beautiful and lyrical love poetry. Complete, unabridged text with explanatory footnotes. 96pp. 5³⁄₁₆ x 8¼. 0-486-27557-4

SANDITON AND THE WATSONS: Austen's Unfinished Novels, Jane Austen. Two tantalizing incomplete stories revisit Austen's customary milieu of courtship and venture into new territory, amid guests at a seaside resort. Both are worth reading for pleasure and study. 112pp. 5⅜ x 8½. 0-486-45793-1

THE SCARLET LETTER, Nathaniel Hawthorne. With stark power and emotional depth, Hawthorne's masterpiece explores sin, guilt, and redemption in a story of adultery in the early days of the Massachusetts Colony. 192pp. 5³⁄₁₆ x 8¼.

0-486-28048-9

THE SEASONS OF AMERICA PAST, Eric Sloane. Seventy-five illustrations depict cider mills and presses, sleds, pumps, stump-pulling equipment, plows, and other elements of America's rural heritage. A section of old recipes and household hints adds additional color. 160pp. 8⅜ x 11. 0-486-44220-9

SELECTED CANTERBURY TALES, Geoffrey Chaucer. Delightful collection includes the General Prologue plus three of the most popular tales: "The Knight's Tale," "The Miller's Prologue and Tale," and "The Wife of Bath's Prologue and Tale." In modern English. 144pp. 5³⁄₁₆ x 8¼. 0-486-28241-4

SELECTED POEMS, Emily Dickinson. Over 100 best-known, best-loved poems by one of America's foremost poets, reprinted from authoritative early editions. No comparable edition at this price. Index of first lines. 64pp. 5³⁄₁₆ x 8¼. 0-486-26466-1

SIDDHARTHA, Hermann Hesse. Classic novel that has inspired generations of seekers. Blending Eastern mysticism and psychoanalysis, Hesse presents a strikingly original view of man and culture and the arduous process of self-discovery, reconciliation, harmony, and peace. 112pp. 5³⁄₁₆ x 8¼. 0-486-40653-9

SKETCHING OUTDOORS, Leonard Richmond. This guide offers beginners step-by-step demonstrations of how to depict clouds, trees, buildings, and other outdoor sights. Explanations of a variety of techniques include shading and constructional drawing. 48pp. 11 x 8¼. 0-486-46922-0